KEYS TO INFINITY

KEYS TO INFINITY

Clifford A. Pickover

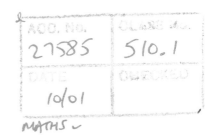

John Wiley & Sons, Inc.
New York ■ Chichester ■ Brisbane ■ Toronto ■ Singapore

Design, composition, editorial services, and production management provided by Professional Book Center, Denver, Colorado.

Unless noted, all illustrations have been created by the author or are reproduced from copyright-free sources. The two-dimensional color computer graphic renditions in the insert were produced using custom C-language programs running on an IBM RISC System/6000 or on an IBM POWER Visualization System. IBM's software Visualization Data Explorer and the Platt Galsoft renderer were used to aid in the production of the three-dimensional computer graphics.

Library of Congress Cataloging-in-Publication Data
Pickover, Clifford A.
    Keys to infinity / Clifford A. Pickover.
       p. cm.
    ISBN 0-471-11857-5 (acid-free paper)
    1. Infinite. I. Title.
    QA9.P515   1996
    793.7'4—dc20   94-45541

Printed in the United States of America

10 9 8 7 6 5 4 3 2 1

This book is dedicated to all those
who do not have a book
dedicated to them.

Acknowledgments

I thank Martin Gardner, J. Clint Sprott, Robert Stong, Arlin Anderson, Clay Fried, and Manfred Schroeder for useful comments. I thank Don Webb for permission to use our collaboration, "Valley of the Horses," and J. Clint Sprott for his permission to use our collaboration, "Escape from Fractalia," in this book.

The angelic engravings interspersed throughout this book come from G. Doré, *The Doré Illustrations for Dante's Divine Comedy* (Dover 1976). Born in Strasbourg in 1832, Gustave Doré was perhaps the most successful illustrator of the nineteenth century. Doré began work on these *Divine Comedy* illustrations in 1857 and usually drew his designs directly onto wood blocks for printing.

The quotes from George Zebrowski come from an article titled "Is Science Rational?" that appeared in the June 1994 issue of *OMNI* magazine (p. 45). Permission to reprint the lyrics from "The Song That Doesn't End" was granted by Norman L. Martin, songwriter.

I have published some of the ideas in this book in the following scholarly and professional journals: "The Loom of Creation" (*Computers and Graphics*), "Fractal Milkshakes" (*Leonardo*), "Automated Computer Art" (*Computers and Graphics*), "Recursive Worlds" (*Dr. Dobb's Journal*), "Random Number Generators" (*The Visual Computer*), "Fractal Batrachions" (*Computers and Graphics*), "Factorions" (*Mathematical Spectrum*), "Undulation of Monks" (*Mathematical Spectrum*), "Vampire Numbers" (*Theta*), "Fractal Curlicues" (*The Visual Computer*), "Carotid-Kundalini Functions" (*Fractal Report Newsletter*), "Logit Terrain" (*IEEE Computer Graphics and Applications*), and "Slides in Hell" (*Skeptical Inquirer*).

Contents

Science is not about control.
It is about cultivating a perpetual
condition of wonder in the face
of something that forever grows
one step richer and subtler
than our latest theory about it.
It is about reverence, not mastery.

—Richard Powers, *The Gold Bug Variations*

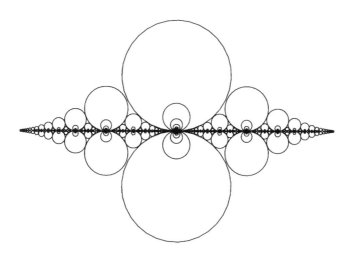

Preface

We live on a placid island of ignorance in the midst of black seas of infinity, and it is not meant that we should voyage far.
  — H. P. Lovecraft

The heavens call to you, and circle about you, displaying to you their eternal splendors, and your eye gazes only to earth.
  — Dante

I could be bounded in a nutshell, and count myself a king of infinite space.
  — Hamlet

I think we become interested in the concept of infinity early in childhood. Perhaps our initial fascination starts when we hear about large numbers, or outer space, or death, or eternity, or God. When I was a boy, I often visited my father's library to examine his large collection of old books. The one that stimulated my early thoughts about infinity was not a mathematics book, nor a book on philosophy, nor one on religion. It was a history book published in 1921 titled *The Story of Mankind*. In the book, Hendrik Willem Van Loon starts with a little parable next to a sketch of a mountain:

> High up in the North in the land called Svithjod, there stands a rock. It is a hundred miles high and a hundred miles wide. Once every thousand years a little bird comes to the rock to sharpen its beak. When the rock has thus been worn away, then a single day of eternity will have gone by.

This idea of eternity—a temporal infinity—is enough to start any child wondering about the inexhaustible fabric of numbers, space, and time.

A few years later, I wandered through my father's library and was rewarded with another book by Van Loon titled *The Arts*, published in 1937. I pulled the dusty book from the shelf and was delighted to find the following philosophical gem, a conversation between a student and a wise, old teacher:

"Master, will you not tell us what the highest purpose may be to which mortal man may aspire?"

A strange light now came into the eyes of Lao-Kung as he lifted himself from his seat. His trembling feet carried him across the room to the spot where stood the one picture that he loved best. It was a blade of grass, for within itself it contained the spirit of every blade of grass that had ever grown since the beginning of time.

"There," the old man said, "is my answer. I have made myself equal of the Gods, for I too have touched the hem of Eternity."

Lao-Kung, like many of the ancient philosophers and writers, considered the concept of God to be intimately intertwined with the infinite. For example, facing this Preface is a view of Heaven from Dante's *Divine Comedy* showing the numbers of angels increasing to infinity the higher one ascends. St. Augustine believed not only that God was infinite, but also that God could think infinite thoughts. According to Augustine, God "knows all numbers." Augustine's works, along with the simple Van Loon quotations, provided a seed in childhood from which my interest in infinity and large numbers grew, and in particular provided an early stimulus for *Keys to Infinity*.

Infinite Worlds

The trouble with integers is that we have examined only the small ones. Maybe all the exciting stuff happens at really big numbers, ones we can't even begin to think about in any very definite way. Our brains have evolved to get us out of the rain, find where the berries are, and keep us from getting killed. Our brains did not evolve to help us grasp really large numbers or to look at things in a hundred thousand dimensions.
— *Ronald Graham*

Prepare yourself for a strange journey as *Keys to Infinity* unlocks the doors of your imagination with thought-provoking mysteries, puzzles, and problems on topics ranging from huge numbers to life itself. Each chapter is a world of paradox and mystery.

My favorite chapter, "Welcome to Worm World," describes the evolution of huge worms on checkerboard worlds. Readers of all ages can study the behavior of these worms with just a pencil and paper. (Growing international interest in this topic has led to a recent short publication in *Discover* magazine.) In "Welcome to Worm World" you'll be among the first to learn about the Internet Superhighway WormWorld Tournament, in which researchers around the world competed to find the longest evolving worms. How does the worms' behavior change as Worm World grows to the size of our universe?

Consider each chapter a launchpad for thinking and experimenting. In "Ladders to Heaven" you are asked to imagine what it would be like to climb an incredibly long ladder stretching from the Earth to the moon. You can use only ropes and other mountain-climbing gear. Impossible, you say? Read further and find out what scientists have to say about such a gargantuan task.

In "The Leviathan Number" you'll learn about a monstrous number so large as to make the number of electrons, protons, and neutrons in the universe pale in comparison. (It also makes a googol—1 followed by 100 zeros—look kind of small.) You'll learn about large numbers beyond the ability of humans to grasp or compute, apocalyptic numbers, superfactorial functions, apocalyptic powers . . . . What *can* we know about numbers too large to compute or imagine?

In "Fractal Milkshakes and Infinite Archery" you'll learn about a bubbly froth lurking in the fabric of our number system. The foam is comprised of an infinite regression of circles known as Ford circles. (A graphical representation of the froth is placed on page ix to whet your appetite.)

In "Slides in Hell" you'll be asked to descend immense porous slides with zany mathematical properties. Want to gamble from which hole in the slide you'll fall? Have you ever dreamed of playing God, simulating life, or preventing cancer? Then the chapter "Creating Life Using the Cancer Game" is for you. Want to fly through immense grids of dots—as big as the universe—with startling properties? Then take a look at "Grid of the Gods." Hop aboard a flying saucer stealing humans from Earth, and compute the sex of the one-billionth abductee in "Alien Abduction Algebra."

Keys to Infinity is for anyone who has pondered the immensity of numbers, dreamed of daring challenges, and wondered about the infinitely small. I hope *Keys to Infinity* will stimulate creative thinking; enhance computer programming skills; and suggest the usefulness of simple mathematics for solving curious, practical, or mind-shattering problems. BASIC and C source programs are included for those of you who own computers. Some of the larger programs are gathered together in Appendix 2.

My Keys

I looked round the trees. The thin net of reality. These trees, this sun. I was
infinitely far from home. The profoundest distances are never
geographical. — John Fowles

To help you on your journey, I offer various keys:

1. Essays on all the previously mentioned topics and more, everything from vampire numbers to the loom of creation.
2. Puzzles, such as the fiendishly difficult cyclotron puzzle, with hints to remind you there are often more ways of looking at the world than are immediately obvious.
3. Quotations from novelists, philosophers, and famous scientists.
4. Program codes, so you can experiment further using personal computers as an aid to your pencil-and-paper explorations.
5. Fractal and other images of infinity to stimulate your imagination. (Fractals are intricately shaped objects that reveal infinite detail as they are continually magnified.)

Some topics in the book may appear to be curiosities, with little practical application or purpose. However, I have found all these experiments to be useful and educational, as have the many students, educators, and scientists who have written to me during the past few years. It is also important to keep in mind that throughout history, experiments, ideas, and conclusions originating in the play of the mind have found striking and unexpected practical applications. I urge you to explore all the topics in this book with this principle in mind.

As in all my previous books, you are encouraged to pick and choose from the smorgasbord of topics. Many chapters are brief and give you just a flavor of an application or method. Often, additional information can be found in the referenced publications. To encourage your involvement, computational hints and recipes for producing some of the computer-drawn figures are provided. For many of you, seeing pseudocode will clarify concepts in ways mere words cannot.

I have created all the compter graphics images in *Keys to Infinity* and have provided a brief description of the color plates. The book chapters are arranged somewhat randomly to retain the playful spirit of the book, and to give you unexpected pleasures. Some of the more technical chapters are placed at the end. Throughout the book are suggested exercises for future experiments and thought, as well as directed reading lists. Some information is repeated so each chapter contains sufficient background information, and you may therefore skip chapters. The basic philosophy of this book is that creative thinking is learned by experimenting.

Perhaps I should say what this book is *not* about. It does not contain the standard number-crunching problems found in scientific texts—most often these do not stimulate creativity, nor do they have artistic appeal. Also, the problems and topics in this book are not of a "linear" variety, where variables are fed into an equation and a succinct answer is returned. In fact, many of the exercises are of the "stop-and-think" variety and can be explored without using a computer.

The book is not intended for mathematicians looking for a formal mathematical treatise. Various books in the past have given fascinating accounts of infinity in mathematics, culture, and art. Eli Maors's *To Infinity and Beyond* and Rudy Rucker's *Infinity and the Mind* describe the history of number theory and various ideas connected to the concept of infinity. Their topics include number series, prime and irrational numbers, Cantor sets, and non-Euclidean geometries. They also discuss infinity in the Kabbalist and Christian concepts of God, as well as astronomers' evolving concepts of the size and structure of the universe. Two other useful books are Stan Gibilisco's *Reaching for Infinity* and Ray Hemmings's and Dick Tahta's *Images of Infinity*. These books, and others, are included in Further Reading at the end of this book.

As there have been so many excellent books on the subject of infinity, *Keys to Infinity* is intended to provide unusual views on the way the human mind makes sense of the world through the use of computer tools, games, puzzles, numbers, and mathematical relations. Many chapters touch on the concept of infinity directly, whereas others are meant to stimulate readers' minds in a more general sense regarding the unlimited extent of time, space, or quantity. I leave more direct discussions of infinity in number theory and culture to my predecessors.

Anti-omniscience

The universe is not only stranger than we imagine, it's stranger than we can imagine. — Arthur C. Clarke

Keys to Infinity emphasizes creativity, fun, and expansion of the mind. For most chapters, no specialized knowledge is required. As I just mentioned, even though many chapters contain mathematical ideas and computer programming hints, almost all problems are of the "stop-and-think" variety that do not require programming or sophisticated mathematics to allow you to explore and imagine.

Many questions I pose in the book are unanswered. Some may be unanswerable. As Stanford psychologist Roger Shepard recently noted at a Sante Fe Institute workshop on the limits of scientific knowledge, even if our computers and mathematical tools continue to improve, we may not understand the world any better. He says, "We may be headed toward a situation where knowledge is too complicated to understand." As Princeton astrophysicist Piet Hut has pointed out, the structure of the physical universe may represent the ultimate limit on human knowledge. John Horgan (*Scientific American*) believes that particle physicists may never be able to test theories that unify gravity and the other forces of nature because the predicted effects become apparent beyond the range of any conceivable experiment.

Finally, Ralph Gomory, a former director of research at IBM, and who is now president of the Alfred P. Sloan Foundation in New York City, believes our educational system does not place enough emphasis on what is unknown or even unknowable. To solve this problem, the Sloan Foundation may start a program on the limits of knowledge. It is hoped that *Keys to Infinity* will stimulate students in thinking about both the unknown and the unknowable.

The Electronic Smorgasbord

There was from the very beginning no need for a struggle between the finite and infinite. The peace we are so eagerly seeking has been there all the time. — D. T. Suzuki

In many chapters of *Keys to Infinity* I quote colleagues from around the world who have responded to my questions, and I thank them for permission to reproduce excerpts from their comments. My questions were sent through electronic mail and often posted to electronic bulletin boards. Two common sources for such information exchange were "rec.puzzles" and "sci.math," which are electronic bulletin boards (or "newsgroups") that are part of a large, worldwide network of interconnected computers called Usenet. (The computers exchange news articles with each other on a voluntary basis.)[1]

Some define Usenet as the set of people (not computers) who exchange puzzles, tips, and news articles tagged with one or more universally recognized labels signifying a particular "newsgroup." There are thousands of newsgroups on topics ranging from bicycles to physics to music. Usenet started out at Duke University around 1980 as a small network of UNIX machines. Today there is no UNIX limitation, and there are versions of the news-exchange programs that run on computers ranging from DOS PCs to mainframes. Most Usenet sites are at universities, research labs, or other academic and commercial institutions. The largest concentrations of Usenet sites outside the United States seem to be in Canada, Europe, Australia, and Japan.[2]

KEYS TO INFINITY

C H A P T E R 1

Euler strode up to Diderot and proclaimed: "Monsieur, $(a + b^n)/n = X$, donc Dieu existe!" ("Sir $(a + b^n)/n = X$, therefore God exists!")
— Michael Guillen, *Bridges to Infinity*

Proof is an idol before which the mathematician tortures himself.
— Sir Arthur Eddington

Too Many Threes

How many numbers contain the digit 3? This sounds like a simple enough question. However, when I first innocently posed this question to colleagues, some nearly fainted when they heard the correct answer. Others, upon hearing the answer, finally realized that a degree in philosophy would be quite valuable in their daily lives—because the thin fabric of reality began to tear just a bit.

Because the answer is quite mind-boggling, some have recommended that it be presented only to those over the age of 18. How many numbers contain the digit 3? The answer is: virtually all of them.

The best way to solve this problem is to actually do it for yourself. Let's start by examining the first ten numbers. In the first ten numbers, 1, 2, 3, 4, 5, 6, 7, 8, 9, 10, there is only one number that contains the digit 3. This means that 1/10, or 10 percent, of the numbers have the number 3, when considering the first ten numbers. In the first 100 numbers the occurrence of numbers with at least one 3 seems to be growing. In fact there are 19 such numbers:

3, 13, 23, 33, 43, 53, 63, 73, 83, 93, 30, 31, 32, 34, 35, 36, 37, 38, and 39

This means 19 percent of the digits contain the number 3 in the first 100 numbers.

We can make a table showing the percentage of numbers with at least one 3 for the first x numbers:

x	Percentage of Occurrence
10	10
100	19
1000	27
10000	34

It is astonishing to most people that the percentages rapidly increase to a mind-numbing 100 percent, indicating that almost all numbers have a 3 in them! In fact, a formula describing the proportion of 3s can be written:

$$1 - (9/10)^n$$

The proportion gets very close to 1 as n (the number of digits in a number) increases.

I can hear some of you screaming, "How can it be that almost all the numbers have a 3 in them?" Brian Kendig from San Fransisco, California, wrote to me: "A three in every number? Preposterous!" Joan Bacurio from Ohio said, "If all numbers have a three in them, I'll eat my hat."

First of all, it's true: Most numbers have a 3 in them. Why? As the numbers get larger, they contain more digits, increasing the probability that one of the digits might be a 3. In fact, the probability that a 3 will *not* appear in a very long number is very low.

Note that the probability that a random sequence of n digits does not contain a 3 is 0.9^n. (I'll explain this later.) As n goes to infinity this probability goes to zero. Seth Breidbart of New York notes, "Since almost all numbers have a lot of digits (there are only a finite number of integers with less than n digits, and infinitely many with greater than n) the limiting probability is 0."

Interestingly, the percentages I show for the first few numbers with at least one 3 also hold for other digits. Given a number with enough digits, there is essentially a 100 percent chance of all digits 0–9 appearing in that number. As a number grows in value, the number of digits in the number also increases. As the number approaches infinity, the number of digits in the number approaches infinity (at a slower rate). As the number of digits approaches infinity, the likelihood of any specific digit appearing at least once approaches 100 percent.

Dan Shoham from Massachusetts Institute of Technology notes that the sequences 15, 172, and 666 (and any other finite sequence) are also contained (in order) within almost all numbers, for similar reasons. John Torrey from North Carolina notes that the probability of a number having *exactly* one 3 is virtually zero for large enough numbers.

Finally, K. Alagarsamy from the Indian Institute of Science constructed the following table that emphasizes the intriguing concept that the number of numbers that contain 3, and the number of numbers that do not contain 3, both reach infinity as one scans larger and larger numbers.

The First n Numbers	The Number of Numbers That Contain the Digit 3 $a(n)$	The Number of Numbers That Do Not Contain the Digit 3 $b(n)$
10	1	9
100	19	81
1000	271	729
.	.	.
.	.	.
.	.	.

Both the sequences $a(n)$ and $b(n)$ are increasing and will reach infinity as n reaches infinity.

Try It Yourself

Mathematicians are looking at life with a most trenchant sense—one that perceives things the other five senses cannot.— Michael Guillen, Bridges to Infinity

The C and BASIC programs given in Appendix 1 for this chapter allow you to compute the proportion of numbers, with and without the digit 3, as more numbers are considered. The programs make use of the formula $1 - (9/10)^n$ as presented earlier in this chapter. Here, n is the number of digits in a number. Table 1.1 lists these proportions for several examples. (See also Figure 1.1.)

TABLE 1.1. Proportion of Numbers without and with the Digit 3

D*	No 3s	Some 3s
0	1.000000	0.000000
1	0.900000	0.100000
2	0.810000	0.190000
3	0.729000	0.271000
4	0.656100	0.343900
5	0.590490	0.409510
6	0.531441	0.468559
7	0.478297	0.521703
8	0.430467	0.569533
9	0.387420	0.612580
10	0.348678	0.651322
11	0.313811	0.686189
12	0.282430	0.717570
13	0.254187	0.745813
14	0.228768	0.771232
15	0.205891	0.794109
16	0.185302	0.814698
17	0.166772	0.833228
18	0.150095	0.849905
19	0.135085	0.864915
20	0.121577	0.878423
.	.	.
.	.	.
.	.	.
100	0.000027	0.999973 (googol)

*D is the number of digits in the number.

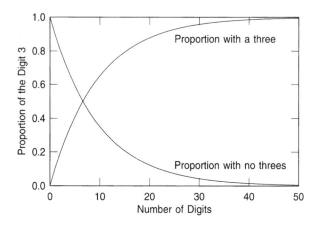

FIGURE 1.1.   The proportion of numbers with and without the digit 3.

Bob Stong from Virginia suggests we create a more interesting program that searches through the integers to find the first place at which the number of numbers containing a digit 3 exceeds the number of those without a 3. I challenge readers to create a program to solve this mystery.

The formula for computing these proportions is best understood as follows. The probability of having a 3 as a digit in a one-digit number is 1/10 and of not having a 3 is 9/10. For a two-digit number, the probability of not having a 3 as the first digit or the second digit (that is, there are no 3s in the two-digit number) is simply the product of not having a 3 for the first digit multiplied by not having a 3 for the second digit: $(9/10) \times (9/10) = 81/100 = 0.81$. The probability of having a digit 3 is $1 - 0.81$. For a three-digit number we have $(9/10) \times (9/10) \times (9/10) = 729/1,000 = 0.729$, and so forth. For an n-digit number we therefore have the probability of not having any 3s: $(9/10)^n$.

The Search for Super-3 Numbers

Three is the most special of all numbers. I know it's a pretty bold statement, but consider the following. Three is the only natural number that is the sum of all preceding numbers. It is the only number that is the sum of all the factorials of the preceding numbers: $3 = 1! + 2!$. In religion, three reigns supreme. In ancient Babylonia there were three main gods: the Sun, Moon, and Venus. In Egypt there were three main gods: Horus, Osiris, and Isis. In Rome there were three main gods: Jupiter, Mars, and Quirinus. For Christians, three symbolizes the Holy Trinity: Father, Son, and Holy Spirit. In classical literature, there were three Fates, three Graces, and three Furies. In languages, there are three genders (masculine, feminine, and neutral) and three degrees of comparison (positive, comparative, and superlative.)

Note that German Chancellor Otto von Bismarck signed three peace treaties, served under three emperors, waged three wars, owned three estates, and had three children. He also organized the union of three countries. His family crest bore the motto *in trinitate fortitudo* ("In trinity, strength"). There is a German saying, *Alle güte Dinge sind Drei* ("All good things come in threes").

"Super-3" numbers are integers i such that, when raised to the power of 3 and then multiplied by 3, contain three consecutive 3s. When I first mentioned super-3 numbers of the form $3i^3$, they existed only in the realm of the mind, as no one had actually discovered such a number. Since then, Walter Pachl from Vienna, Austria, discovered the first super-3. It is 261, because $3 \times 261^3 = 53{,}338{,}743$. Are there others? Are they distributed randomly through the integers, or is there some sort of pattern? Walter Pachl used the following REXX program to find what is believed to be the smallest super-3 number, 261:

```
/**/
Parse Arg a
If a='' Then a=1
Numeric Digits 1000000
Do I = a By 1
   x = 3*(i**3)
   If i//10=0 Then Say i length(x)
   If pos('333',x) > 0 Then Do
     Say 'I found one: i='i
     Call lineout 'super3.res',i x
   End
```

As this book goes to press, Arlin Anderson from Alabama has become the world's most proficient super-3 hunter. He has discovered, for example, that $3 \times 471^3 = 313{,}461{,}333$, and he believes that *every* number ending in the digits 471, 4710, or 47100 is a super-3. For example, he also discovered that $3 \times 1{,}471^3 = 9{,}549{,}030{,}333$.

What's special about the years 1923, 1926, and 1968? It turns out that these are all super-3 years because they are super-3 numbers. Will humanity pass through another super-3 year in your lifetime?

Regarding the presence of super-3s in our number system, Arlin Anderson notes:

Numbers larger than 10^{500}, when cubed, have at least 1,500 digits, with a high probability of 333 occurring somewhere in their string of digits. Therefore, almost all numbers greater than 10^{500} will be super-3 numbers. (There will, however, be an infinite number of exceptions, of course! For instance, 1, 10, 100, 1,000, 10,000, . . . are never super-3 numbers.)

Finally, is the following statement true or false? "There is an infinity of numbers that do not contain the digit 3." Does this conflict with the statement, "Almost all numbers contain the digit 3"?

Ladders to Heaven

That's the problem with eternity, there's no telling when it will end.
— Tom Stoppard, *Rosencrantz and Guildenstern Are Dead*

Imagine what it would be like to try to climb an incredibly long ladder stretching from the Earth to the moon. Impossible, you say? Read on to learn more about a recent question I posed to a number of scientists, engineers, and computer programmers. For those of you who like to program computers, Appendix 1 contains some BASIC and C code to carry out similar (safer) experiments in the privacy of your own home. For those who prefer to philosophize without a computer, there is ample material here for lively debate and group discussion.

Consider the following scenario. A standard ladder stretches from each country on the Earth upward a distance equal to the distance from the Earth to the moon. I asked scientists to assume the following:

1. The ladder is made of a strong metal, such as titanium, that will not break.
2. The ladder is inclined at a very steep angle, 70 degrees, for each country.
3. There is a breathable atmosphere.
4. The people (or teams of people) are allowed to use standard mountain climbing and camping gear (such as ropes and backpacks) but not sophisticated electrical mechanisms such as engines.
5. A $1 million reward is given to whomever reaches the top of the ladder first. In addition, the country that reaches the top first has its national debt wiped out.

Next, I posed a list of questions to ponder:

1. Approximately what length of time would be required for a person (or team of people) to reach the top of the ladder: Days? Weeks? Years?
2. Which country would be the first?
3. Is there any novel method you would suggest to reach the top of the ladder?
4. Is this task impossible to carry out?

Before describing some slightly crazy, yet ingenious, mechanisms for carrying out this task, I'd like to mention a few observations people made. For example, some

FIGURE 2.1.   Will any humans actually see a view like this from the ladder to the moon, or would it be impossible to travel the necessary distance?

said it was not clear whether the moon (Figure 2.1) were in the proposed scenario or whether the Earth were assumed to be alone in the universe for the sake of simplicity. Spyros Potamianos of Hewlett-Packard was the first to suggest using the moon's gravitational field to "fall upward" once one had reached the point where the moon's gravity is greater than the Earth's. He also asked whether the climbers must survive the fall, or whether simply having a dead explorer reach the top of the ladder is sufficient for a country to win the prize.

Ken Arromdee from Johns Hopkins University suggested that the country that gets to the top fastest would be the one closest to the Earth's equator, where the centrifugal force is the greatest. Others noted that countries with a large national debt have extra motivation.

Life on Ladder World

This section includes a variety of suggestions from colleagues. How many of these ideas do you think would work?

Kenneth Tolman from the University of Utah believed that a million dollars was not a significant incentive for climbing a ladder 3.5×10^8 meters long. He noted that, for best results, there would be a large crew, some leading and some providing support, food, and other necessities. Ropes could be arranged in a pulley system. At the ends of each pulley, people would attach and detach packages. Waste packages would be sent down from above by attaching them to the top of a particular rope pulley system. At the same time, a new package would be attached at the bottom. The person running the pulley would compensate for any slight difference in the weight by pulling the presumably slightly heavier new packages upward. A possible problem with this scenario is that the strength of ropes or cables may be insufficient. In addition, the friction of the system will be very large.

Kenneth estimated that a person in good physical condition could climb 10,000 feet in a day. (He personally has climbed this distance in a single day while exploring the Grand Canyon.) Kenneth also assumed that a professional climber would be three times as good as he, and therefore can travel 10,000 meters a day. This requires 3.4×10^4 days for the job, or at least 93 years. Kenneth said, "We will have to have teams trade off, because the first teams will be dead by the time they get to the top. Pregnant women will have to climb along, too, to provide new climbers."

The task of providing food and supplies is not trivial, and Kenneth considered this to be the major bottleneck in the operation. He calculated that if one person were positioned every 10 kilometers, this still required a support-staff population of 30,000 to feed. The support staff could be drastically reduced by allowing air support to provide food. People might grow food on different levels in communities that then could function to pass food and needed supplies upward.

The task of feeding 30,000 people distributed at 10-kilometer intervals over a thin ladder is difficult to imagine. There obviously would be a major distribution problem. If each person ate only 0.06 pounds in a day, a ton of food would have to be passed upward each day. However, it would be impossible for a single human to move a ton of food 10 kilometers upward in a single day. Kenneth therefore concluded that the task requires communities living at different heights, collecting rain water and growing their own food. However, he strongly believed that the task would never be undertaken, even with the rewards mentioned, "unless it were a command from God."

Matt Crawford also suggested some ingenious ways for completing the long journey. He would make use of a bicycle-like vehicle clamped to the ladder. By pulling a light, strong rope on a pulley, riders could be changed fairly quickly if they made use of a "crew of brawny pulley-pullers." Variable-geared linkage to the rope would help. Matt noted, "It seems impossible for the rider to pull this ever-longer rope, but I think shorter segments could be lifted and linked. Or the ground crew could help the rider by pulling down rope from a hub of lesser diameter than the wheels of the vehicle."

Matt did not believe the task is impossible to carry out. He first thought it might be impossible to stop the movement of the vehicle at the far end of the ladder and return, due to centrifugal acceleration, but that acceleration turns out to be only about 5 cm/s^2.

How do we get food to the people? Matt suggested the riders change so often that they only need high-carbohydrate snacks and several quarts of fluid. "I think the brawny ground crew could pull up the next rider (with supplies and another pulley and segment of rope) at an acceleration of 0.5 g or better." This would require less than 90 minutes for each shift-change up to the synchronous orbit level. Again, frictional forces may pose the greatest problem to actually implementing this.

Montgomery Graf from the Island Graphics Corporation suggested that one would be able to fashion a hot air balloon given condition 4. Also, given condition 3, the hot air balloon would be able to cover the entire distance. One would then only need to attach a sliding hookup between the ladder and the balloon, and wait.

Jeff Woods from CompuTrac, Inc. (Texas), suggested human or animal power to hoist a human using one big pulley at the end of the ladder, but there would be the problem of how to first position the pulley at the end of the ladder. He next suggested having a string of people cooperating to push each other along with some mechanism that could increase the distance between people as they went. For example, climbers could push a jack on the ladder, then lift it to insert another jack, and repeat the process while each jack is being extended. In this way, the leading edge would accelerate constantly to the end of the ladder.

Jeff considered the bicycle idea (mentioned previously) to be the best idea if it could be combined with a gear system. Instead of a rider pedaling and requiring a massive support team trying to feed itself and the rider, Jeff suggested using a vehicle powered by a gear with a diameter smaller than the wheels. The cable is pulled from the ground by "lots of humans being paid at minimum wage." The vehicle is propelled with no increase of mass on the ladder. The only increase would be in the cable itself. He says, "Nonperishable supplies could be attached to the cable being fed out. Hopefully, the vehicle and cables won't wear out." Could humans build a sufficiently strong cable?

Tom Rankin of Poughkeepsie said, "If my calculations are correct, the end of the ladder would be hurtling through space at about 60,000 miles per hour with respect to the moon." (This is a typical speed for meteors traveling through space.) Depending on the arrangement of ladders in the scenario, this might mean that a person (or the ladder) would crash violently into the moon, making quite a large crater. Of course, people on the ladder would not feel the rotational movement of the ladder if the atmosphere in which they operated rotated with the Earth. Similarly, we don't currently feel the rush of the Earth around its axis at about 700 miles per hour.

Dr. J. Clint Sprott, a physics professor at the University of Wisconsin, wondered whether a lone individual had any hope of climbing such a ladder. He says:

> If I can climb a 10,000-foot mountain in a day, my average power output is sufficient to overcome the gravitational potential of the Earth in about six years. Since I couldn't carry more than a tiny fraction of the provisions for such a journey, I concluded that it was impossible to escape without help. The atmosphere, and its accompanying wind resistance, would also be a major detriment since I would want to put most of my power output into kinetic rather than potential energy during the intermediate stages to cover the distance in only six years. To get to the moon in six years, I'd have to average almost five miles an hour.

Finally, Arlin Anderson from Alabama suggests that we build X-shaped machines with a wheel on each tip. Each wheel is ratcheted so it cannot roll backward. The bars of the X can slide against one another so the rider uses a natural climbing or sliding motion to run the machine. At distances greater than 3,000 km, gravity will be less than a half-g, so the rider can give a big "kick" and coast for a few seconds. (His last push will be at about 37,000 km. At this point he might expire, but his dead carcass would reach the moon, and he could therefore win the prize.)

Arlin believes it would be reasonable to assume an average speed of 130 km/day. If possible, he suggests making the air cold so the rider does not sweat. In addition, he suggests that a giant, flat water bottle with a small food cylinder be mounted along the main axis of the X. The food and water system would weigh about 50 lbs. and would offer sustenance for two weeks. Helpers would push food up the ladder and use a similar X-shaped go-cart. All the carts would be hooked together, and whenever the food and water in a cart were not sufficient, the person behind the depleted rider would pump food and water up through tubes to the rider. After the first day, the helpers in back are already feeling hungry and thirsty because they have pumped about half their load forward. However, their lighter load makes their job easier, and they continue to ascend. After a week, the last quarter of the train is falling back to Earth, the next quarter is now on half-load, and the front half is still happy. After the second week, the second quarter drops out and the third quarter is on half-load. After the third week, the third quarter drops out, but the fourth is still loaded. In three more weeks, the cycle is repeated, and the train will be one-sixteenth as long as when it started. After 18 weeks, it will be 4,095/4,096 as long. What is left (one cart, one rider) is fully loaded! This is made possible because the rider had 4,095 helpers. Because the rider will not be hungry when he hits geosynchronicity, it is likely that slightly fewer than 4,095 friends would be necessary, "unless we demand he be alive when he splats the moon."

Insight from Computer Programs

BASIC and C code are provided in Appendix 1 so you can compute the weight of a human as he or she ascends the ladder. This gives rise to many unexpected insights, which I discuss shortly. In this scenario we assume the presence of both the moon and the Earth.

The necessary equation to compute the ratio (W) of your weight (at a distance d_1 from the center of the Earth) to your usual weight on Earth involves the masses of the Earth (M_e) and moon (M_m) as well as the radius of the Earth:

$$W = (M_e/(d_1^2) - M_m/(d_2^2)) \times (R_e^2)/M_e$$

Here, d_2 is the distance of your body to the center of the moon. The lower curve in Figure 2.2 shows the ratio of your new weight to your normal weight as you ascend the ladder. Obviously, the ratio is close to one when your distance from Earth is small (left side of graph). (For simplicity, we are ignoring the fact that an Earth-attached ladder also has centrifugal forces to consider.)

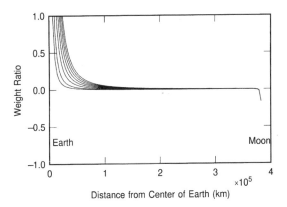

Distance from Center of Earth (km)

FIGURE 2.2.   The ratio of your new weight to your normal weight as you ascend the ladder (bottom curve). The second curve from the bottom was computed using a hypothetical Earth with twice its usual mass. The next curve uses four times the mass; the next, six, and so on.

Most people I've queried intuitively thought that as a climber ascended, he or she would continue to gradually lose weight until some magical point between the Earth and the moon were reached, at which point the climber would be weightless, and then the climber would gain weight and start plunging into the moon. In reality, the weight-ratio function has a broad plateau at $W \sim 0$, which means that for the bulk of the trip, one would essentially be weightless (see lower curve in Figure 2.2). The effect of the moon is so tiny that only in the very, very end of the trip would the effect be measurable. This effect is indicated by the small negative blip at the right of the graph. (The curve goes negative because the explorer is now being pulled in an opposite direction.) On the moon, an explorer would weigh about one-sixth of what he or she does on the Earth.

How would this curve change if the Earth were more massive than it really is? The other curves in Figure 2.2 indicate the effect of changing the mass of the Earth. The second curve from the bottom was computed using an Earth with twice its mass. The next curve uses four times the mass; the next, six; and so on. Even with an Earth 14 times its actual mass, the traveller would spend most of his or her time essentially weightless.

The Infinite Ladder

Throughout this chapter, we have not considered whether or not such a ladder could be built. Most physicists and engineers suggested that the ladder would be impossible to build with current materials because the Earth's gravitational force would snap the ladder before it reached a height of 100 miles. Some engineers, such as Alrin Anderson, suggested that the centrifugal force produced by the spinning Earth would weaken the ladder more than the force of gravity. To strengthen the ladder out in space, Arlin suggests that the cross-sections of the ladder sides be doubled every 100 or so miles for strength. One problem with this idea is that a 1-inch-diameter ladder side would grow as large as Jupiter by the time the ladder were 23,000 miles high (the point where the Earth's centrifugal force balances gravity). In the future, stronger materials may make the ladder easier to build.

Digressions

Here are some additional questions for you to ponder. What would Figure 2.2 look like if the moon were replaced by the sun? You should be able to determine this using the program codes provided in Appendix 1. The necessary astronomical constants are:

1.98×10^{30} Mass of sun (kg)
696265   Radius of sun (m)
1.49×10^{11} Mean Earth-to-sun distance (km)

Would the journey be made more difficult with the gravitational effect of a third planetary body placed midway between the Earth and the moon? Where is the center of mass for the Earth-moon system? Arlin Anderson notes that the moon's mass is about 1/80 of the Earth's, and the moon is almost 240,000 miles away. Therefore, the center of mass of the Earth-moon system is 3,000 miles from the center of the Earth, or 1,000 miles *below* the Earth's surface! I am interested in hearing from readers with additional insight into how humans could efficiently climb the long ladder postulated in this problem.

Finally, as this chapter goes to press, I've come across the fascinating book *What If the Moon Didn't Exist? Voyages to Earths That Might Have Been*, by Neil Comins (HarperCollins, 1993). How would life on Earth be different today if the moon never existed? Comins says the Earth would rotate more than three times faster than it currently does. There would be recurrent gale-force winds like those on Jupiter, making it a challenge for "tall" beings (such as humans) to evolve. Our atmosphere would have taken millions of years longer to convert from one dominated by carbon dioxide to the oxygen-rich atmosphere of today.

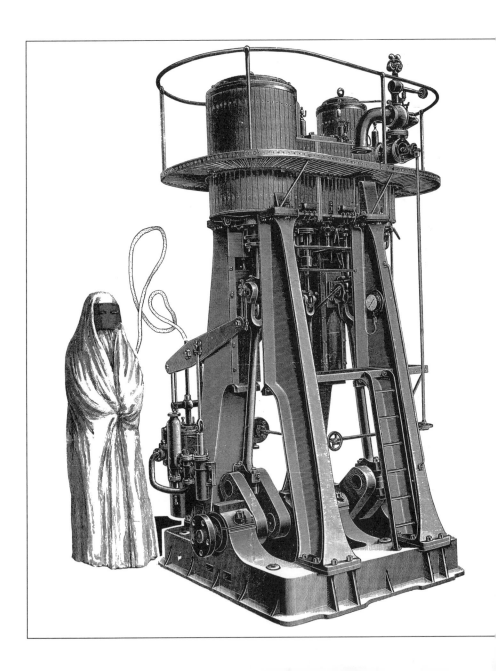

CHAPTER 3

Infinity Machines

Infinity commonly inspires feelings of awe, futility, and fear.
— Rudy Rucker

I like infinity. I believe that infinity is just another name for mother nature. Nature provides infinite possibilities all the time. But because we have suffered through this world of wars and woes, we sometimes fail to get this. We see the world as a little stingy at times.
— Fred Wolf, *Parallel Universes*

My interest in strange mechanical devices soared to new heights when I stumbled across the conference proceedings from the International Tesla Society.[1] The society generally discusses untested (and sometimes untestable) technological ideas and bizarre, fascinating inventions. The Tesla Society derives its name from world-class inventor Nikola Tesla (1856–1943) who was born in Serbia and immigrated to the United States in 1884 to work with Thomas Edison. Tesla invented the induction motor and polyphase power transmission, both still of considerable importance today. He was responsible for the first practical commercial use of alternating current (AC) motors (Figures 3.1 and 3.2), generators, and transmission lines. Tesla also developed a high-voltage generator known the Tesla coil, which to this day is used to give spectacular lightning demonstrations. (Perhaps you've seen demonstrations in museums where streamers of electricity shoot from a large metal ball.)

Despite Tesla's real technological genius, he did advance a panoply of eccentric ideas. For example, Tesla wanted to pump electrical energy into the atmosphere of Earth so people would be able to run appliances without plugging them in. He also was working on inventions that he thought would make the atmosphere fluoresce, so there would be no nighttime on Earth!

Perhaps even crazier than these ideas was his eccentric personality. Tesla had many fears, such as fear of germs and fear of round objects, particularly pearls on women's necklaces and earrings. He could not eat a meal without first counting all the objects on his plate. All "countable" food items, such as beans, french-fries, or crackers, had to come in multiples of three or he would not eat them.

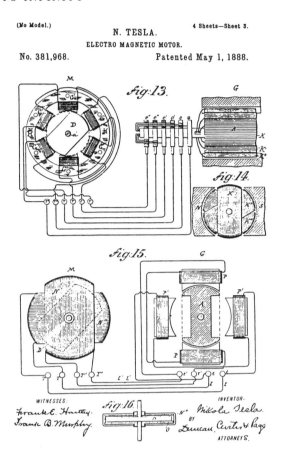

(No Model.)   4 Sheets—Sheet 3.

N. TESLA.

ELECTRO MAGNETIC MOTOR.

No. 381,968.   Patented May 1, 1888.

FIGURE 3.1.   Nikola Tesla's patent for an electromagnetic induction motor.

Despite these eccentricities and sometimes dubious inventions (Figure 3.2), Tesla was a genius—and the International Tesla Society probably overlooks some of his wilder "inventions." On the other hand, the society's workshops are often concerned with recent attempts at perpetual motion machines, machines that seem to produce more energy than they consume, and machines that seem to extract energy from some unknown physical realm. For example, Joseph Newman, an attendee of their conference, spent 14 years fighting unsuccessfully with the U.S. Patent Office for recognition of his "perpetual motion energy machine." You can read much more about the society's unusual devices in Jeff Johnson's *Skeptical Inquirer* article (see "For Further Reading" at the end of this book).

This discussion about strange inventions and perpetual motion machines naturally leads to the topic of this chapter. For many years, I have been interested in "infinity machines"—devices that rely on the concept of infinity for their operation. Although the devices may seem fanciful, their philosophical underpinnings are profound. Can you detect a flaw in the principle of their operation? Would these de-

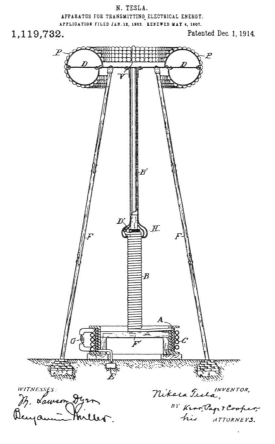

N. TESLA.
APPARATUS FOR TRANSMITTING ELECTRICAL ENERGY.
APPLICATION FILED JAN. 18, 1902. RENEWED MAY 4, 1907.

1,119,732.   Patented Dec. 1, 1914.

FIGURE 3.2.   Nikola Tesla's patent for transmitting electrical energy through the air.

vices work in a theoretical sense, if we were not limited by the physical strength of the materials needed to build them?

Infinity Antilock Brakes

Human beings know a lot of things, some of which are true, and apply them. When we like the results, we call it wisdom. — Herbert Simon

Many traditional automobile brakes work by converting your foot's pressure on a brake pedal to force applied to brake shoes that push against the inner surface of a brake drum attached to the car's wheel. The newer antilock brake system (ABS) functions similarly, except that a steady foot pressure is converted to a rapid on/off pushing of the brake shoes against the drum. This pulsing action reduces the possibility of the shoes locking against the drum, and thereby reduces skidding of the wheel on slippery pavement. The pulsing rate is generally a constant 4 to 15 times per second depending on the manufacturer.

Infinity antilock brake system (IABS), designed quite recently, looks like any other antilock brake system. The pulsing rate of the shoes against the drum, however, is not constant but rather increases in speed as long as your foot is applied to the brake pedal. The system is guaranteed to stop any car in an efficient manner. (When I inquired about the IABS at my Mitshbishi 3000 dealer, I was told that this device was not yet available.)

The increase in the rate of pulsing is as follows. When you apply force to the brake pedal, the shoes press against the drum for 1/2 second, then the shoes move away and release pressure for 1/4 second, then they press again for 1/8 second, then they are off for 1/16 second, and so on. It turns out that this infinite series (1/2 + 1/4 + 1/8 + . . .) adds up to 1. Therefore, at the end of 1 second, the brake shoes have switched an infinite number of times because 1 second = 1/2 second + 1/4 second + 1/8 second . . . .

Question: Are the brake shoes on or off the drum at the end of the second? For a number of physical reasons, it would be difficult to actually design an infinity antilock brake system, but let's discuss this in a theoretical manner for the moment. You seem to have all the necessary information to determine whether the brake shoes are pressing on the drum or not. Additionally, the shoes have to be either pressing or not pressing. What is your answer?

Infinity Keyboard

Some of you are probably arguing that the rapid pulsing required in the infinity antilock brake system just described may cause an overheating of the brake drum and shoes, despite the use of cooling fins bonded to the outside of the brake drum to increase the rate of heat transfer to the air. Vents in the wheels to increase air circulation for cooling also may not reduce temperatures sufficiently. Therefore, I introduce a less mechanically stressed system, the infinity keyboard.

Imagine that an extremely agile (supernatural) being alternately presses the J key and the H key on your personal computer (PC). The being presses the J key for 1/2 second, then it presses the H key for 1/4 second, then the J key for 1/8 second, then the H key for 1/16 second, and so on. Again, this infinite series (1/2 + 1/4 + 1/8 + . . .) adds up to 1. Therefore, at the end of 1 second, the keys have switched an infinite number of times.

Question: What is the last letter typed at the end of the second? You seem to have all the necessary information to determine whether the last letter is a J or an H. What is your answer?

Infinity Program

Finally, we can eliminate all the heavy mechanical work (if you consider pressing keys heavy mechanical work) by designing an infinity computer program. The following program computes the same infinite series (1/2 + 1/4 + 1/8 + ...), and prints the first few results.

```
/* Infinity C Program */
#include <math.h>
#include <stdio.h>
int i, result;
float r = 1, sum;
main()
{
    i=1; sum=0;
    while (sum<1) {
        r = pow(0.5,i); sum=sum+r;
        if(i<5) printf("Result is: %d %f\n",result,sum);
        i++; if((i%2)==0) result=1; else result=0;
    }
    printf("Final Result is: %d %f/n",result,sum);
}
```

With every step in the infinite series, the variable "result" is toggled between a 1 and a 0 value. Here are the first results:

Result	Sum
0	.5
1	.75
0	.875
1	.9375

As we've mentioned, the sum of the series will eventually be equal to 1, at which point we jump out of the "while" loop and print the final result. Is the final result 1 or 0? You seem to have all the necessary information to determine whether the last value of the variable is 1 or 0, even without running the program. What is your answer?

Some Closing History of Infinity

The concept of infinity has challenged humans for centuries. For example, Zeno (an Eleatic philosopher living in the fifth century B.C.E.) posed a famous paradox involving infinity. The paradox seemed to imply that you can never leave the room you are in. As Zeno reasoned, to reach the door you must first travel half the distance there. Once you get to the halfway point, you must still traverse the remaining distance. You need to continue to half the remaining distance. The procedure can be repeated as diagrammed on the following page:

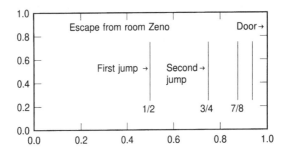

If you were to jump one-half the distance, then one-quarter the distance, then one-eighth the distance, and so on, will you reach the door? Not in a finite number of jumps! In fact, if you kept jumping forever at a rate of 1 jump per second until you are out the door, you will jump forever. Mathematically one can represent this limit of an infinite sequence of actions as the sum of the series $(1/2 + 1/4 + 1/8 + \ldots)$. The modern tendency is to resolve Zeno's paradox by insisting that the sum of this infinite series $1/2 + 1/4 + 1/8 \ldots$ is *equal* to 1. Because each step is done in half as much time, the actual time to complete the infinite series is no different from the real time required to leave the room. (One easy way to compute the sum of this series is to use the formula $y = 1 - 2^{-n}$, which gives the sum of the first n terms. For $n = 10$, $y = 0.99902$.)

Many Greeks, however, could not accept the existence of infinity. In fact, when one of Pythagoras' disciples, Hippasus, found that $\sqrt{2}$ was irrational[2] they killed him, for his colleagues could not accept this infinitely long, nonrepeating number. However, other Greek philosophers, when learning that the square root of 2 was not a rational number, celebrated the discovery by sacrificing 100 oxen. (Weren't humans more passionate about mathematics in those days?)

Paradoxes Past the Speed of Light

Various wonderful paradoxes arise with infinity machines. In this section, a few of these paradoxes are highlighted by a variety of readers.

Steve Frye from North Carolina suggests that the probability of finding the brake engaged is 2/3. For example, the first time the shoes press against the drum they press for 1/2 second, then they move away and release pressure for 1/4 second. The brake continues to spend twice as much of the remaining time engaged as disengaged.

However, paradoxically you can also examine the terms shifted by one. When examined this way, does the brake also appear to spend twice as much of the remaining time disengaged? Is the probability of finding the brake disengaged also 2/3?

Next, Steve considers that if the brakes are to engage and disengage in shorter and shorter times, the brakes will soon "be moving at warp speed (and believe me, the brakes will warp), which is somewhat higher than the traditional light-speed barrier." To counteract this seemingly paradoxical situation where the brakes are moving faster than the speed of light, we can move the brake shoes closer to the drum with each braking action, so the distance the shoes need to travel is shorter, and the switching will take place at below light speed. Unfortunately, at some point, the distance that the brake shoes move must soon become a minute fraction of the radius of a quark. Thus, the brake will effectively (if not philosophically) be engaged and disengaged and in the same state. Because this process proceeds inexorably, we must conclude from this discussion that the probability of ending up both engaged and disengaged is 1.

Another interesting fact is that if the brakes are to arrive at a final state of engaged or disengaged, there must be a final switch. But this is in direct conflict with the fact that infinity goes on forever; there is no last term in the series. So the brakes cannot be in either one or the other state; they must be in both states or neither state. The discussion above argues strongly for "both engaged and disengaged."

Tom Randall from New York also had some fascinating comments on my infinity machines. Many of his comments echo the concerns of Steve Frye. Tom notes that the brake and key-press examples involve the motion of an object having mass. As the distance traveled by these parts remains constant, while the time allowable for the motion decreases, one would encounter a series of mechanical problems relating to the acceleration rates and speeds necessary to accomplish the required motion in the time allotted. Assuming one managed to overcome all these problems, one would eventually encounter (Einsteinian) relativity problems. In other words, like Steve Frye, Tom Randall suggests that the parts would need to attain a speed greater than the speed of light to move the required distances in the required time. Tom notes these infinity machines would stop running before 1 second had passed.

Can we calculate what length of time the infinity antilock brakes or infinity keyboard would operate before this relativistic effect was seen? To answer this question, we would need to know the distance travelled by the brake shoes or the key, and we would need to assume infinite energy available to accelerate the parts. We also would need infinitely strong, heat-resistant materials to make the parts.

Of course, the C program given earlier in this chapter would also have some problems if run on any available machine using any available compiler. The problems would involve the finite precision available for the floating point numbers used. Eventually the program will be trying to add a value to the sum that must be rounded either to zero or to a higher value, which will cause the sum to be greater than 1. Thus the program will stop earlier or later than the 1-second mark, and its last output value will not represent conditions at exactly the 1-second mark.

C H A P T E R 4

Maps are art with a purpose.

— Lloyd Brown

Give me a map; then let me see how much is left for me to conquer all the world.

— Christopher Marlowe, *Tamburlaine*

Infinity World

Fifth Avenue

One day while taking a stroll down Fifth Avenue in New York City I noticed a street vendor selling maps of the Earth. One colorful map was placed above the other, stretching up along the side of a skyscraper for about 15 feet. (The repetition of identical patterns reminded me of the works of Andy Warhol.) We will probably never know whether the vertical display of maps produced additional sales for the street vendor, but the scene stimulated me to conduct the "Infinity World" survey in May 1994 in which I asked scientists, sociologists, cartographers, and other interested parties to speculate on a hypothetical Earth formed by the infinite repetition of vertically placed maps. What would civilization be like on an Earth of infinite area? Would there be fewer wars? How would global politics be affected? Before launching into this discussion, I'd like to provide some background to the history of mapmaking.

Ancient Maps

Not one of the lands written into your destiny will speak to you the
language of your first atlas. — Primo Levi

In ancient Western civilizations, the world was considered to be ridiculously small. The Earth was generally believed to be shaped like a flat pancake floating on the surface of the infinitely deep, surrounding ocean. This disc was large enough to hold all the land known at that time: the shores of the Mediterranean Sea, with adjacent parts of Europe, Africa, and a bit of Asia (Figure 4.1). According to Aristotelian cosmology, the entire universe was small, and space was thought to be finite and to have a definite edge (Figure 4.2).

Of course, humanity's knowledge of Earth's geography gradually improved through time. However, even in 1544, maps of the Earth contained some striking anomalies. One favorite example is Battista Agnese's map of the world. Though in many respects accurate, it still portrayed North America as an impressionistic blur, a borderless figment of the cartographer's imagination. Interestingly, even as late as the 1740s, maps often depicted California as an island.

FIGURE 4.1.   The world of the ancients of Earth. The Earth was believed to be a large, flat disc floating on the surface of the world ocean that surrounded it. Reprinted by permission from *One, Two, Three . . . Infinity*, by George Gamow. (New York: Dover, 1988).

FIGURE 4.2.   Finite space. According to Aristotelian cosmology, space was finite and had a definite edge. This idea was accepted during medieval times on Earth. The illustration here is often said to be a 16th-century German woodcut, although its origin is probably much more recent.

Good maps have always accelerated the pace of civilization. For example, in the fifteenth century the Turks closed lucrative trade routes from the Orient to Europe. This motivated the development of new routes, primarily by sea, to restore mercantile connections with the Orient. These new routes and avenues of trade, in turn, accelerated cartographic and navigational technologies. As IBM scientist Dan Platt has pointed out, these two applied sciences, one of them visual and the other analytical, worked hand in hand to allow humans to circumnavigate the globe for the first time. These sciences allowed Europeans to dominate the globe in less than a century. For the first time, the Earth became a global community. The power of merging visual and analytical technologies transformed humanity.

Worlds without End

Those who explore an unknown world are travelers without a map; the map is the result of exploration. The position of their destination is not known to them, and the direct path that leads to it is not yet made.
— *Hideki Yukawa*

In this chapter, I focus on the concept of Infinity Worlds: hypothetical Earths with infinite extent. For example, consider the geography of the Earth today, except that the map is infinitely repeated in the north and south directions (Figure 4.3). On Infinity World one could fly north in a plane from Greenland and arrive at the South America of the adjacent map.

How would life and geopolitics be different today if humans had evolved on such an infinite Earth? When I asked colleagues this question, I indicated that I was more interested in the sociology, psychology, religion, and politics of such a world than in opinions on plate tectonics, gravity, and pure biophysical effects, although all ideas were welcome. For the sake of discussion, colleagues were allowed to assume that all physical laws on our Earth are maintained on Infinity World. When respondents asked for more information, I suggested that the infinite repetition of "maps" might be thought of as being on the surface of a cylinder.

I also asked colleagues how their answers would be different if Infinity World were created by repeating today's world map along a vertical line in the Atlantic or Pacific ocean. (In these scenarios the world would be infinitely repeated horizontally rather than vertically.) If they thought it useful, respondents were told they could think of a repetitious map on a extremely wide cylinder or sphere. Scientists were asked to ignore the fact that gravity would flatten humans in a superwide Earth.

Finally, I also asked the following questions that referred to a single Earth model, and not to an infinitely repetitious map:

1. How would the world be different today, geopolitically speaking, if the ancient land masses had never drifted apart and, therefore, today's world consisted of a single supercontinent? How would biological life be affected?

2. What would today's world be like if the land mass that formed the Greek peninsula never existed? Would this effect be lesser or greater than if the Italian peninsula never existed?

FIGURE 4.3. A small section of Infinity World created by an infinite repetition of Earth maps. (Schematic representation by JoAnn McLoughlin.)

3. About 15,000 years ago humans crossed a land bridge that joined Asia to Alaska at the time. What would today's world be like if that land bridge never existed?
4. Why do all the major peninsulas on Earth point south? Consider, for example, Italy; Greece; Florida; Baja; and the tips of Africa, South America, India, Norway, Sweden, Greenland, and many other land masses.

Vertical Infinity Earths

The map is the game board upon which human destinies are played out, where winning or losing determines the survival of ideas, cultures, and sometimes entire civilizations. — Stephen Hall, Mapping the Next Millennium

The profound effects of an infinitely repetitious world are not entirely clear. Would wars be infrequent because territorial acquisition is unimportant in an infinite world? Would humanity's awesome sense of wonder in Infinity World decrease the occurrence of wars? In my opinion, wars would be just as frequent. Wars were quite common when our Earth was largely unexplored and the Earth had "unlimited" territory. Ancient civilizations were often at war.

Would space travel evolve on Infinity World? Because there would always be unexplored terrestrial regions, why would the inhabitants look to the stars for adventure, national prestige, or scientific advancement?

Would ecological concerns and "green" political movements evolve on Infinity World? Why would inhabitants be concerned with pollution, biodiversity, or animal or plant extinction with unlimited land and water? No doubt there would be local pollution concerns, but overall interest would be diminished.

Would an organization resembling the United Nations evolve? Would the dinosaurs still be alive today on Infinity World? Would criminals be more difficult to catch? Instead of fleeing to South America, they would only have to hop a plane to any one of a number of "maps." On Infinity World, there are infinitely many countries in which to hide.

In my opinion, religion would be profoundly affected because the existence of identical, repetitive land masses would suggest the presence of a creator as opposed to the random assortment of lands that we now have. (Can you imagine what effect the existence of a perfectly square or circular continent would have on religion and science?)

Sociopolitical Impact

In this section, I turn our attention to comments I received from other scientists and interested readers. Some colleagues, such as David McAuley from Trinity College Dublin, suggested that the term *superpower* would become meaningless on Infinity World. How could a nation capable of any significant multiworld regulation evolve? Given an infinite number of communities, all kinds of bizarre sociopolitical and religious organizations would evolve. Certainly some would have extremely violent dispositions, which would lead to anarchy in parts of Infinity World. Planet-wide communication, of course, would be impossible, and humans would not know the "full" extent of the world.

Craig Becker from Austin, Texas, wondered whether intelligent life would develop on only *one* of the maps, or on all of them at roughly the same time. If life evolved on just a few maps, the consumption of material resources would be of less concern. If human cultures developed independently on a few hundred maps, there still would not be significant movement among maps until the twentieth century because of the inhospitable polar regions. These cold boundary regions would tend to isolate one world from another before air flight developed. In fact, most respondents noted that humans would have to await the production of airplanes before they discovered the worlds beyond the Antarctic barrier. Even jet travel would only be of limited use. A commercial plane's unrefueled range is less than 15,000 miles. As a result, some respondents suggested that trans-Arctic railroads would move bulk goods past the Antarctic barrier between airports between worlds.

Mike Hocker from Poughkeepsie, New York, suggested that the use of atomic weapons on Infinity World would be less of a concern. Mike believes that wars would be common as populations tried to destroy others before they were destroyed. Xenocide and paranoia would be the geopolitical norm. Nations would race to develop interworld ballistic missiles (IWBMs) as opposed to intercontinental ballistic missiles (ICBMs). Vast land areas would be radioactive or devastated by horrendous nuclear, chemical, and biological weaponry.

Mike Hocker echoed my own opinions when he suggested that "resource depletion" would take on a new meaning, as would species extinction. For example, if Pacific Ocean whales are in abundance in oceans 30 worlds away, inhabitants of one world can pollute their own Pacific Ocean with less concern. Of course, the humans living 30 worlds away may have the same view and be polluting their own Pacific Ocean, too.

On Infinity World, disaffected persons can "leave" their world and go to a new place. This is not possible on our limited Earth at this time, and this absence of a safety valve permitting exodus causes stagnation and societal stratification. (Consider European stagnation and persecution before the New World was discovered and accessible.)

Alain Vaillancourt from Montreal, Canada, speculated on mythological and theological effects of a repetitious Earth. He says, "Would people postulate a stuttering Christian god who created Earth on a really nervous day, repeating it ad infinitum?"

Jim McLean from Boca Raton, Florida, suggested that all Earths would evolve identically if started from the same initial conditions. For example, if all parallel Earths started at a completely identical state, divergence of events would not occur, and every dewdrop on every blade of grass would be identical on each sub-Earth. Your exact twin would exist on each and every sub-Earth, but you could never meet your twins because for every step you take north your twin also steps north.

Chaos theory, which suggests that imperceptibly small differences are amplified through time, may have no effect on Infinity World if all sub-Earths have truly identical starting points. McLean suggests we allow some uncertainty to exist in our starting points. Perhaps we will let electron positions be random, and then chaos will eventually diversify our sub-Earths.

Even if we let small initial differences between Earths cause a divergence of events, your nearly identical twin would still exist, because there are infinitely many Earths. Note, however, that the number of sub-Earths containing your identical twin, though infinite, is vanishingly small next to the total number of sub-Earths where your twin does not exist.[1]

We've all heard the cliché about an infinite number of monkeys placed in front of an infinite number of typewriters, with one typing the complete works of Shakespeare by chance. Jim McLean suggests that each sub-Earth is like a very complex monkey in front of a vast typewriter. Every possible sequence of events, no matter how unlikely, will occur in an infinitely repetitious world. This infinity effect has interesting consequences. What would be the psychological implications of knowing you have an infinite number of twins, or of meeting your twins? Some of your twins would be virtually identical. Others would be slightly different, whereas others would have died at various ages or become mass murderers. On Infinity World, even unlikely possibilities become plausible.

From a mathematical standpoint, McLean notes that there is little danger of actually running into your parallel-world doppelgänger, as the number of sub-Earths where you are not born so vastly outweighs the ones where you are born. (Still, an infinite number of your equivalents exist on an Infinity World.) And, although the infinity effect means that some extremely bizarre sub-Earths will exist, some version of the central limit theorem (the statistical notion that repeated combinations of random observations tend to cluster around a single mean) will cause Earths to cluster around some average behavior.

McLean believes that wars on Infinity World would be fought between people far enough away to be considered "different" but close enough for logistical practicality. (This pattern holds in our own historical development.) Initially there would be wars between clans and villages, later between countries, and then between continents. As technology advances, wars will be fought between sub-Earths, and finally between latitudinal sub-Earth alliances against their cross-polar neighbors. In addition, McLean believes that a large fraction of Infinity World is likely to regress centuries due to ballistic nuclear confrontations across polar boundaries. As communication and transportation methods improve, important individuals (a sub-Earth's equivalent of Hitler or Gandhi) would affect adjacent worlds. (What would happen if Gandhi-1 met Gandhi-2, and they collaborated to spread their message? What if Hitler-1 and Hitler-2 collaborated?)

After the wars between Earths, information-age technology would begin to link surviving sub-Earth neighbors. Distances between sub-Earths would shrink, and the pace of cultural cross-fertilization among sub-Earths would increase dramatically. As on our own Earth, the overall pace of technological and scientific advancement would be set by the most advanced societies, but those societies would not necessarily raise the standard of living significantly among poorer neighbors. Eventually, large clusters of sub-Earths would share common cultures, mores, and gene pools. We would all become swirls in an infinite melting pot.

Some respondents assumed that civilization starts on only one Earth map. If this happened, explorers would immediately know what to expect on other Earths. They would know the areas to avoid because of bad climates, floods, droughts, or earthquakes. Once an earthquake occurred on the San Andreas fault of Map 1, explorers would not settle in the same place on Map 2.

Would confusion reign on Infinity World because a sovereign nation owns different territories in different maps? For example, the government of Vietnam might occupy southeastern Asia on one world and North America on the next.

Curtis Karnow, a partner at the San Francisco law firm of Landels, Ripley, and Diamond, sums up his feelings eloquently:

We do, of course, *already* live in an infinite world: the universe. But life is lived locally. Local politics, nearby neighbors, the restaurant around the corner, the girl next door . . . . For all of our travelling and movement in this country, I'll bet the vast majority still live locally, and the rest want to. The locale may change of course, from time to time, but the human eye focuses best around 15 feet; the long-distance stare is sightless: It looks not out, but in. The stars are for the occasional sighting; we do not, really, steer by them these days. This is not to deny that we need that sense of limitlessness and infinite possibility, just there peripherally, out of the corner of the eye, as remarkable as oxygen . . .

Physical Constraints to Infinity World

A map of the world that does not include Utopia is not worth even glancing at. — Oscar Wilde

Of course there would be various physical problems associated with Infinity World. David Karr from Cornell University in Ithaca, New York, wondered how such a huge, infinitely long object could orbit the sun. He suggested that the sun might orbit the Earth, and therefore Galileo could not have challenged the Church by claiming, "The Earth moves!" (Would there be seasons on an infinite Earth? On our Earth, there are seasons due to the tilt of the Earth with respect to the plane of the Earth-sun orbit.)

From strictly thermal considerations, only a finite number of the Earth "maps" would be habitable because regions far from the sun would be too cold. There might exist distant sections made habitable by other orbiting stars, but these would have no effect on the history of our map or its near neighbors.

If the sun's orbit were a helix with an extremely low pitch, the sun might travel very slowly down the helix, and all living things could follow the sun's trajectory as they migrated beneath the sun's warmth. (I like to imagine a double helical spiral, like DNA, in which dual suns could orbit the cylinder.) In a single-sun world with the sun in a helical orbit, very few fossils would be available for scientific studies, because the panorama of evolution would be spatial as well as temporal. Fossils would be deposited in regions of Infinity World millions of miles away from current life and the current position of the sun and hence would be unavailable for discovery. Perhaps the theory of evolution would never have emerged, and hence religious opposition to evolution would never have occurred.

David DeLaney from the University of Tennessee's Physics Department suggested that Infinity World would not have a magnetic field, if the cylinder were hollow. This in turn would have extensive effects on navigation, the Van Allen belts, cosmic radiation, animals' navigation, auroras, and so on.

Brian Pickrell assumed the polar regions to be stretched out into a rectangular map projection prior to fitting the Earths on a cylinder. If the Earth were an infinite tube, like a giant paper-towel roll, gravity would decrease with distance as $1/r$, rather than $1/r^2$. Normal satellite orbits would be impossible. Pickrell suggests that there could not be any planets or stars. He also believes that an infinite number of suns, moving in some gravity-defying pattern, would be required to illuminate the entire world. Because there would be no Coriolis force making winds, the weather would be completely different. There would be no high- and low-pressure areas, hurricanes, or trade winds. Instead, large convection patterns would create winds that never change.

Alain Vaillancourt from Montreal, Canada, had similar concerns. He suggested an infinite tubular sun illuminating the infinite Earth cylinder, or a series of suns spaced at regular intervals along the tube. The tube might be perfectly straight, or another possibility is that the universe would be comprised of an infinite number of spaghetti-shaped planets. Alain suggests that on Infinity World an immortal human (or a cruise ship bearing thousands of families) could walk from one end of the galaxy to another. This would be of interest to astronomers.

On Infinity World, land animals (including most birds) would be completely cut off from crossing into other worlds. With millions of years of separate evolution, vastly different biotas would exist.

Alaska-Asia Land Bridge

What would today's world be like if the prehistoric land bridge that joined Alaska to Asia never existed? David Karr suggested that the Americas would have been populated by way of Hawaii, resulting in an aboriginal population with a different racial and cultural flavor. However, these North Americans would have suffered the same fate as the American Indians when the Europeans arrived.

Mike Hocker believes that the absence of a prehistoric land bridge would have a huge effect on our world. Eohippus, a primitive horse, was present in both North America and Europe (54,000 years ago), but the principal development of the horse occurred in Europe. The absence of a prehistoric land bridge would probably imply that no horses would be available to Europeans. For some unknown reason, horses became extinct in the Americas while their European children lived on. Would there be any horses today without a land bridge? If no horses existed, the impact would be tremendous because of the role they played in human transportation and as farm labor.

Others also noted that the ancestral Indians and Eskimos could have come to the Americas by boat. In support of this position, note that people reached Australia perhaps 35,000 years ago without a land bridge. However, if the New World were uninhabited when Europeans arrived, the differences in the United States would be moderate, but Latin America might never have been settled by the Spanish and Portuguese. Historically, the explorers were more interested in conquering than in colonizing. In such a case, Central and South America would have been settled by the English, French, and Dutch.

Peninsular Anisotropy

Why do all the major peninsulas on Earth point south? (Consider Italy; Greece; Florida; Baja; and the tips of Africa, South America, India, Norway, Sweden, Greenland, and many other land masses.)

I was surprised by the number of individuals who did not agree that most of the major peninsulas point south. One respondent suggested that many major peninsulas point north, but that these are located in inhospitable places such as Antarctica and are therefore overlooked.

Ken Arromdee from Johns Hopkins University suggested that because most of the land on Earth is concentrated in the Northern Hemisphere there is a higher probability for peninsulas to point south than north. (Incidentally, all peninsulas on Antarctica point north.)

Simon Bradshaw suggested the splinter shape of southern continents results from the breakup of Gondwanaland, an ancient supercontient, centered on a "fracture point" at the former junction of the southern tips of South Africa, South America, and India. As for the smaller (noncontinental) peninsulas, Simon believes that any south-pointing bias is by chance. He also remarked that the statistical bias depends on how the term *peninsula* is defined.

One Supercontinent

How would the world be different today, geopolitically speaking, if the ancient land masses had never drifted apart and, therefore, today's world consisted of a single supercontintent? How would biological life be affected?

My thoughts are that the diversity of languages would be far less in such a world. Linguists such as Johanna Nichols from the University of California at Berkeley have done extensive studies reconstructing the spread of prehistoric languages based on comparative linguistics. Languages multiply more rapidly in tropical areas along coastlines and more slowly in the drier interior of continents. The island of New Guinea harbors 80 families of languages, the greatest density of languages found anywhere in the world. On the other hand, a much larger region such as Australia contains only about 30 families of languages. If the land masses of our world never divided, language diversity would be much less than we have today on our Earth.

David McAuley suggested that because climatic conditions would be altered, there would be a massive desert in the center of the supercontinent that would have very extreme seasons.

Steven Bradshaw from the United Kingdom suggested that most trade would be accomplished by inshore sailing, as there would be little motivation for "blue-water" voyages. Technologies driven by the need to cross oceans would be developed more slowly. (Consider the decreased need for long-distance sailing, navigation, steamships, and undersea cables.) Sea-power empires, such as the British Empire, would not develop.

If the supercontient never broke up, there would be no totally isolated biomes. Therefore, disparate species such the Australian marsupials or the Old World and New World primates would not have evolved.

Some readers pointed out that the world's land masses have actually rejoined and separated many times in the past. The supercontinent we are familiar with existed "only" 200 million years ago and was just the most recent supercontinent.

If the continents never separated, Brian Pickrell suggested that the geological effects would cause unpredictable differences in the existence of oil and metal ores. Trade would be much more difficult and politically restricted, as most countries would have no seacoast and therefore wouldn't be able to send their ships out on the high seas. Trade barriers and tariffs would be much harder to circumvent.

In general, there would probably be large deserts and much less fertile land due to the decreased coastlines. Because Antarctica would be joined with the rest of the world, we would live on an Earth with more accessible land. Historical progress would probably have been faster, due to the increased contacts among civilizations. There would be less diversity. Biological evolution would be slower, due to the reduced environmental variation and less change over geological time.

Neil Dullaway of Stockholm, Sweden, suggested the possibility that the entire world would have been united by a common culture very early in its history. Individuals such as Alexander the Great could easily have conquered the entire world.

No Greece or Italy

Some respondents I queried suggested that the world would not be affected to a great extent by the removal of the Greek or the Italian peninsula. In other words, civilizations such as the Greeks and Romans would have been formed anyway, just in another location. The ancient Hellenic tribes (c. 1000 B.C.E.) would have moved into some other area, such as Asia Minor. Presumably, they could have had a similar historical effect.

Neil Dullaway of Stockholm, Sweden, suggests that one of the main heritages from the Greek and Italian peninsulas has been language. Most Western languages are based on both Latin and Greek. If either of these languages had not existed, modern languages also would not exist in their present form. Languages would be based either on those of the Eastern Mediterranean cultures or on the original languages existing in the historical sea powers: Britain, Portugal, and Spain.

Neil also suggests that ancient Greek math was sometimes a hindrance to the development of modern science. Intellects such as Newton finally broke free from axioms inherited from the Greeks. Therefore, Neil feels that we may actually be better off without the Greek influence. Also, consider that Islam has been the source of much modern mathematics. The numerals we use in arithmetic and the concept of zero probably came from Islamic scholars. (There was some controversy among readers on this topic. Arlin Anderson suggests that the Arabic manuscript that introduced numeration to Europe was a translation of a Hindi one and that all of modern trigonometry is based on the Hindu angle-based functions rather than the Greek chord-based ones.)

The Italians had a profound effect on religion. Christianity, for example, gained momentum from the Roman Empire. In Neil's opinion, without the Italian peninsula, we would not have Christianity. Because Islam was an early offshoot of Christianity, there also would be no Islam. Neil speculates, however, that monotheism has been an important factor in modern cultural development, so some kind of Jewish offshoot similar to Christianity would have developed, but this would have depended heavily on the political scene of the Mediterranean at the time. In Neil's opinion, the Italian peninsula has had a greater effect on today's civilization than the Greek peninsula.

Reader Questions

Some readers, such as Daniel Winarski from Tucson, Arizona, were prompted to ask many questions of their own. Would intelligent life have evolved if our atmosphere had been denser? Would more animals fly? If the atmosphere were less dense, would humans have never flown? How does the density of the atmosphere impact the sociopolitics of our world? What would life be like on an Earth with half as much gravity or twice as much gravity? What would life be like on an Earth with one-fifth as much gravity or five times as much gravity?

Consider an Earth where certain elements (such as radium, uranium, or lead) were never present in the Earth's crust. Would humanity have been better or worse off in the long run? If our day were 200 hours long, would we nap more? What would we do if our day were 1 hour long? What if we had a ring instead of a moon? What if Earth were a "flatland"—exactly the same as our Earth, but utterly flat? What if there were no mountains to cause isolated areas, no impenetrable forests to stimulate the imagination, no seasons to keep us appreciative of change, or no nighttime because we had two suns?

Don Webb, a science-fiction author from Texas, notes that *The Girls in the Golden Atom* (1924), by Ray Cummings, describes infinitely small universes and infinitely large worlds. Cummings was inspired by a Quaker Oats box that showed an endless series of slightly smaller labels decreasing into nothingness.

Brian Pickrell asked why most U.S. states and most countries of the world are oriented east-west. Neil Dullaway of Stockholm, stimulated by my question on the removal of the Greek or the Italian peninsula, asked: What would be the effect on today's world if the British Isles had never existed? Would we all speak French or Spanish? What if the Mediterranean Basin (the "cradle" of western culture) had not existed? What if Africa had never existed? Suppose the land bridge joining Africa to the rest of the world had not existed.

Horizontal Repetition on Infinity World

Travel on an Infinity World made from a *horizontal* repetition of Earths would be easier than on a north-south repetition. However, based on our history, Craig Becker suggested that humans would not have explored beyond the first few adjacent maps until modern times due to difficulties in traversing the large distances during the eighteenth and nineteenth centuries. He believes that even with today's level of technology, there would be only a few scattered colonies on the second or third most removed maps—assuming civilization evolved on just one map.

Brian Pickrell suggested that some nomadic tribes would follow a "head east forever" strategy, always moving into virgin wilderness. In each generation, some people would stay while others moved on. The frontier people would eventually be overtaken when a civilization closer to the center developed airplanes and started moving outward. The Antarctic barrier prevents this in the north-south Infinity World. After humanity spread in an ever-widening frontier, Craig believes that the most civilized and populated area would be concentrated near the "parent world."

Simon Bradshaw from the United Kingdom focused his response on how an advanced civilization might *build* a horizontal Infinity World. If one were to create a huge ring (similar to Ring World, created by science-fiction writer Larry Niven), 12,000 miles pole to pole, one could squeeze 50,000 Earth maps along the inner circumference of the ring. Assuming Niven's Ring World dimensions (a 1-million-mile wide strip), one could wrap repetitious maps around and around in a spiral. Assuming a map is 25,000 miles wide and allowing for a 25,000-mile "guard band" between Earth wraps, one could wrap worlds 20 times, yielding a strip containing 1 million Earths. Although not infinite in length, this is certainly large and physically realizable, in theory. Climate would be globally uniform.

Hemicentrism

The reason all modern maps show the northern continents on the top is because the great mapmaking civilizations developed in the Northern Hemisphere. Why would they want to put all the world they knew at the bottom? (This bias is called ethnocentric geography or hemicentrism.) If civilization in the Southern Hemisphere had earlier grown in size and sophistication, and developed mapmaking skills, then today south would be at the top of maps, and compass needles would point south instead of north. Barry Evans, author of the fascinating book *Everyday Wonders*, further suggests that the names of the constellations would be different. For example, viewed from Australia, Orion the Hunter is standing on his head—hardly a heroic pose. The direction clockwise would be opposite from how we now define it. (Our clocks rotate as does the shadow on a sundial in the Northern Hemisphere. In the Southern Hemisphere, the sundial's shadow moves counterclockwise.) Barry Evans also speculates that the art of navigation would have developed more slowly in the Southern Hemisphere because there is no equivalent to Polaris, the North Star, in the southern skies. Polaris appears to stand still, while other stars rotate around it, because it is almost exactly aligned along the celestial north pole. This property is useful to navigators, as the star always lies in the direction of north.

Artificial Earths

For many years, I have been designing artificial Earths and pondering the results of such Earths on humanity's worldview. In my book *Mazes for the Mind*, I described an amusing computer game that I called Cro-Magnon Conquest. The game has since provided hours of intrigue for students interested in modelling the spread of human civilizations. First the computer draws a map of the world. If this is too difficult, you can represent each of the continents by square or rectangular regions. Next select a site for the origin of humans. One possible location is in East Africa. From this point on the map, have the computer use a random number generator to move your humans on the map of the world and see where they move in a random direction, for each increment of time. This is known to mathematicians as a random walk. For example, if your original screen coordinates are at (x_0, y_0), change them by adding ± 100 miles to x and y for each time step. If you don't have access to a computer, why not use dice to control the movement of your humans on graph paper?

As your humans, represented by a dot on the screen, move about, why not have the computer draw a line so you can see the trail your humans leave? By doing this, you will gradually be drawing a crinkly line that soon meanders all over Africa. Whenever it hits the edge of a continent, simply reflect the line back (rather than have it wander off into an ocean).

How many time increments, or "years," does it take for your human tribes to migrate to Egypt and then out of Africa into the Tigris-Euphrates Valley? As mentioned earlier, about 15,000 years ago humans crossed a land bridge that joined Asia to Alaska. How long does it take your humans to arrive in North America and South America, by randomly walking on the map of the world?

Spiral Earths

For those of you interested in other strange map projections, the book *Spiral Symmetry* has figures showing the Earth transformed to snail-like spirals, doughnuts, cubes, and dodecahedrons. My favorite, the map of the Earth warped to a spiral snail shell, was computed by Agnes Denes, a New York artist. She has also mapped the world to lemons (prolate ovoids) and hot-dog shapes. For further information, see Hargittai and Pickover, *Spiral Symmetry*.[2]

Mecca Maps and Heart-Shaped Maps

Martin Gardner, in *Time Travel, and Other Mathematical Bewilderments* (1988), also discusses unusual map projections. The "Mecca map" is designed to quickly indicate to a Moslem the exact direction he must face when praying at any spot on the globe. One way to draw such a map is to make a stereographic projection with Mecca at the plane's tangent point. A cardioid (heart-shaped) map, invented by Johann Werner, was popular during the sixteenth century. B. J. S. Cahill of Oakland, California, patented his butterfly map in 1913. In this representation, the world is projected onto eight triangular faces.

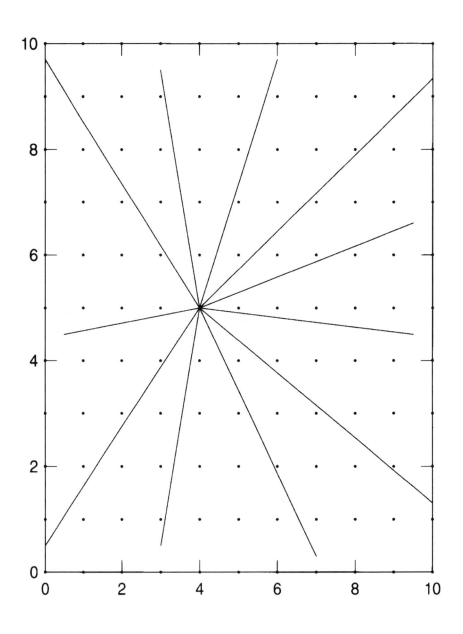

Grid of the Gods

The universe: a device contrived for the perpetual astonishment of astronomers.
— Arthur C. Clarke

I thought of a labyrinth of labyrinths, of one sinuous spreading labyrinth that would encompass the past and the future and in some way involve the stars.
— Jorge Luis Borges

About a year ago, I wrote a science-fiction story about an astronomer named Kalinda. While gazing out into space using a powerful telescope, she saw the most startling arrangement of stars beyond the constellation called Canis Major. She immediately picked up her phone and called her friend, a Dr. Carl Sagan, for advice and counsel. Both Kalinda and Sagan were stumped by the peculiar arrangement. What Kalinda discovered with her new, powerful telescope was a perfectly arranged array of stars in the shape of a cubical grid. About a hundred stars formed each edge of the cube.

To confirm her initial observations, she decided to use the Hubble space telescope's new wide-field camera, which more clearly resolved the star grid, some 7 billion light years from Earth. Because peering at distant stars is like looking back in time, the images revealed a star grid that existed when the universe was 60 percent of its current age.

A torrent of questions flooded Kalinda's mind. How could she make sure she would be the first to publish a research paper on this exciting phenomenon? Was this a problem of national security? Could such an arrangement of stars be a natural phenomenon? What would the tabloid magazines do with this information, once they found out? Would religions be affected or be born? Would she have the opportunity to appear on the TV show *Hard Copy* with her incredible find?

This fanciful story led me to pose the "Grid of the Gods" scenario to students and researchers. I asked colleagues to consider the same sort of grid as in my science-fiction story. The exact dimensions are as follows. A grid of infinitesimal dots

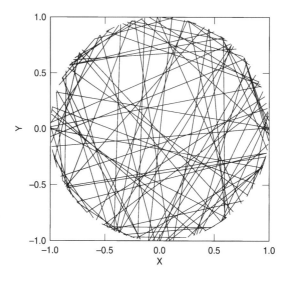

FIGURE 5.1.   One hundred randomly selected chords in a circle.

are spaced 1 millimeter apart in a gargantuan cube having an edge equal in length to the diameter of the sun. For conceptual purposes, you can think of the dots as having unit spacing, being precisely placed at 1.00000 . . . , 2.00000 . . . , 3.00000 . . . , and so on. When I use the term *infinitesimal*, I mean to imply that the dots are not bulbous objects that cover a range of locations in space. They really are located exactly at 1.00000 . . . , 2.00000 . . . , and so on, and do not have a thickness making them extend to, for example, 1.010000 . . . , 2.010000 . . . , and so on.

If you like, you can make a diagram on paper to help you carry out the next step of the problem. At the opening of this chapter is a schematic illustration of a piece of the front face of the cube. It looks like a checkerboard with a dot at each corner of the squares. First pick a dot, any dot, in this densely packed cube of dots. After you have made your selection, draw a straight line through the dot and extend it from that dot to the edge of the cube in both directions.

Assuming infinitesimal dots, is it possible to draw a line starting on one of the dots and extending to the edge of the cube without touching another dot? Does the shape of the grid matter? For example, if a hexagonal grid were used would your answer change? What is the probability that your line will intersect another dot in the fine grid of dots within the cube the size of the sun? Would your answer be different if the cube were the size of the solar system?

The answer to these questions might shock you. Essentially it is certain that all lines through your initially selected dot will never meet another dot! (This assumes that you choose the direction of all the lines randomly. If, for example, the direction of a line were chosen precisely parallel to one of the cube's edges, then the probability of hitting another dot would be 1.)

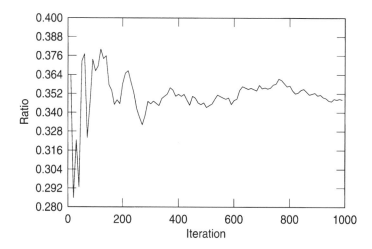

FIGURE 5.2.   The ratio of the number of chords shorter than the circle's radius to the total number of chords. As the simulation proceeds, the ratio approaches 0.33.

Could a computer program be written to simulate this process? Most colleagues I asked said "no." Simple programs could never simulate infinitesimal points. On the other hand, even with the limited floating point precision available on most computers, you can still perform the experiment of randomly placing random lines on a large grid, and you'll find that virtually none will intersect the grid. With infinite precision placement of the points, no line would ever intersect the grid!

Perhaps even more exciting is the following simulation involving the laying down of random lines. I call the simulation the "toothpick" simulation because it reminds me of tossing toothpicks onto the ground. About a year ago, Fermilab scientist Matt Crawford asked, "What is the probability that a randomly selected chord of a circle is shorter than the radius of that circle?" (A chord is a line joining any two points on the circle.) Most scientists I asked could not readily give me the answer to this problem, so I resorted to the BASIC and C code included in Appendix 1 to perform a simulation. To solve the mystery, the programs randomly select chords and determine how many of them are shorter than the radius of the circle. Figure 5.1 shows 100 randomly selected chords. Figure 5.2 shows the ratio of the number of chords shorter than the circle's radius to the total number of chords that the program computed. As the simulation proceeds, the ratio seems to be converging on a value near 0.352 after 1,000 trials. After 100,000 trials, the ratio is around 0.33. Therefore, we can say that one-third of all the possible chords of a circle are less than the circle's radius. If you were a gambling person, you should not bet that a randomly selected chord was smaller than the circle's radius![1]

Interestingly, Fran Asbeck from Endicott, New York, believes it is possible to solve the Crawford question in an analytical fashion using simple geometric diagrams. Here is his note to me:

Although the programs that perform the simulation are interesting for a number of reasons, I believe it is possible to arrive at the one-third probability as follows. Consider a circle centered at C. Choose a point on the circle, Q. Construct chords PQ and QR such that they are equal in length to the radius CQ. Just to make it obvious, construct line segments $CP, CQ,$ and CR, forming two equilateral triangles CPQ and CQR. From geometry, we know that these two adjacent triangles fill 120 degrees of the circle's sweep, leaving 240 degrees of the circle remaining. Any random point on the circle, X, will enjoy a one-third probability of falling within the 120 degrees and a two-thirds probability of falling without. Chord QX then has a one-third probability of being shorter than the radius.

It turns out that one way of looking at the Grid of the Gods problem is by using Cantor set theory. Georg Cantor (1845–1918) was interested in the kinds of problems posed in this chapter. For example, on a quite related topic, he showed that there is an infinity of rational numbers (terminating or recurring decimals such as 0.666666 . . . , 0.5, and 0.272727 . . .) and an infinity of irrational numbers (nonterminating, nonrepeating decimals such as π and $\sqrt{2}$). Interestingly, as discussed in other chapters of this book, Cantor proved that the infinity of irrationals is "infinitely greater" than the infinity of rationals. This gives a hint as to why we can place infinitely many lines between the precisely placed grid of points.

Paul A. Kuckein, an electrical engineer from California, writes the following, which may be of interest to mathematically inclined readers:

While driving home after thinking about your chapter on Grid of the Gods, I considered an infinite universe made up of stars at fixed grid points. Such a universe would have a number of stars that could be mapped, one by one, onto the set of integers; hence, the number of stars would be of (what mathematicians call) Cantor's infinite order $\aleph_0$ The number of lines from a given star to each other star would be infinite, also of order $\aleph_0$ What is interesting to contemplate is that the number of lines from *all* stars to *all other* stars can also be mapped onto the set of integers with the appropriate algorithm. Thus, the infinity of lines from one star to all others is "equal" to the infinity of all lines from all stars to all others. But if you consider the number of lines through a given star that do *not* pass through any other star, the number of such lines is infinite of the order $\aleph_1$ Therefore, in my infinite universe, there are an infinite number of lines from all stars to all others, but there are *more* lines from each star that *do not* pass through any other star!

Digressions

God not only places dice, but He also sometimes throws the dice where
they cannot be seen. — Stephen Hawking

Is it possible to consider the same sorts of questions posed in the Grid of the Gods assuming the lines you draw have a finite thickness? Can you modify the simulation codes to answer the following unanswered questions? What is the probability that a randomly selected chord of a square is shorter than the side of the square? If you were a gambling person, would you bet that a randomly selected cord in a triangle is smaller than all the edge lengths of the triangle? What would be your answer for a sphere or a tesseract (a four-dimensional cube)?

Strahlkörper

The kinds of problems discussed in this chapter are related to what mathematicians call visibility functions and what mathematician Hermann Minkowski called *Strahlkörper* (ray bodies). For more on information on *Strahlkörper* see Schroeder, *Number Theory in Science and Communication*.[2]

CHAPTER 6

In principle . . . [the Mandelbrot Set] could have been discovered as soon as men learned to count. But even if they never grew tired, and never made a mistake, all the human beings who have ever existed would not have sufficed to do the elementary arithmetic required to produce a Mandelbrot Set of quite modest magnification.

— Arthur C. Clarke, *The Ghost from the Grand Banks*

To the Valley of the Sea Horses

If one needed data to support the truism that money cannot buy happiness, one would need only look to the life of Herman Poole.[1] At 65, just after the end of his seventh marriage and the beginning of his eighth, he was worth 16 million dollars and could not name a single week in his life as a happy one.

But he did try.

He did look.

He went to Robofest mainly looking for a new toy or, at best, a new place to play at being a philanthropist. When he arrived on a warm, breezy afternoon, the Austin Coliseum echoed with children's laughter, the whir of gears, and the bubbling and bellowing of strange music. Bright orange extension cords held in place by silver duct tape crisscrossed the floor. Fragrant mixtures of cotton candy, hot dogs, and the latest pheromone colognes filled the air.

Herman Poole's expression was inscrutable as he gazed at all the commotion. Silver Mylar blimps dive-bombed the long tables, and everywhere computer screens flickered. A robot hand crawled by, chased by a Mylar balloon shaped like a man. A man in a synth suit danced, and his motions made music. Amidst all this noise, color, and space, Herman felt small and lost. *I'm like this place*, he thought, *a dismembered part of me here, part of me there. My life is many small compartments, most of them unhappy.*

He wandered out of the circus noise of the main hall and into a smaller, quieter area. To Herman's right, a pretty, young woman in pastel shorts demonstrated the latest in teledildonics. He thought she was a bit too obvious with her short shorts and perfectly manicured hair and nails. As she talked she kicked at the disembodied robot hand, which had somehow wandered into her booth. It scurried away to the left, where Herman saw other booths with book vendors, robot sculptures, Internet demonstrations, and virtual cars.

Near the corner stood a slightly soiled sign marked "Fractal Tours" in bright red. Beneath the sign was a PC on a long, battered table usually used for city garage sales, and on the monitor was a beautiful image. A geometrical pattern of some kind, it mesmerized Herman. An excited young black man was talking about the pattern. Four other people sat on rusty, dented folding chairs listening to the lecture. Herman sat on an empty chair and began to listen.

The young man pointed at the monitor. "That beautiful image, ladies and gentlemen, is a Mandelbrot set." Herman noticed that the man had said the last two words with a certain degree of reverence, almost awe. The young man went on to explain that the shape was called a fractal because of its geometric intricacy. Amazingly, it was produced by a very simple formula, $z = z^2 + c$.

The man seemed to be looking directly at Herman. "It's a mathematical feedback loop," the man said. He stopped, as though he were so impressed with what he had just said that he had to take a deep breath. "The computer sticks in one value for z, and the equation squares it, adds a constant, and produces a new value. This new value is stuck back into the equation, and the process repeated hundreds of times." He paused again, taking another deep breath. "When you square the usual real numbers, nothing interesting happens. Stick in a 2, and you get 4. Stick in the 4 and you get 16. Larger and larger and larger. For complex numbers, however, the story's very different."

"Complex?" Herman said, surprised at himself for interrupting the man's demonstration.

The young man stopped suddenly and pointed at Poole. "Didn't they teach you anything in geometry class?"

Herman was silent. The young man sighed, himself a little embarrassed at having alienated a member of his audience. "Complex numbers," the young man said, "have a real and an imaginary part. When you stick complex numbers into the formula, they sometimes get larger and larger as they speed off like demons to infinity. For other complex numbers, the iteration, or repetition, of the formula doesn't produce exploding behavior."

"How do you get the colors?" Herman asked, pointing to the display.

"That's the easy part," the man said. "The rates at which the numbers speed off to infinity are represented as colors. For example, you can use bright colors to represent the fastest points."

Herman realized that even though he didn't quite understand the explanation, the mathematics somehow produced a pattern so complex that humans couldn't have imagined such beauty, such detail, before computers came along with their fast number-crunching abilities. The patterns reminded Herman of something Jackson Pollock might have done while high on LSD.

The other people watching the display and listening to the young man's pitch seemed less impressed than Herman. A barefoot young woman with dishwater blonde hair got up and left. As she passed by, Herman noticed she smelled faintly of marijuana. One of the other people, a slightly paunchy, bald businessman, interrupted the demonstration to ask about what kind of computer platform it was running on. Could it also run on a Mac? Could you print the pattern? The questions destroyed the beauty of the young man's spiel. The other two people left, as did the bald guy, who probably asked similar questions of all the demonstrators—perhaps to show he felt himself clever enough to be part of this subculture.

Herman noticed that the young man lecturing on fractals wore a laminated "Demonstration" card on his denim jacket. The young man's name was Malcolm Caswell. The letters of his name were printed in an ornate, bushy-looking font.

Herman said, "When I walked up, you had the Mandelbrot set on display. Now you have all these spirals. They're pretty. What are they?"

"That's a close-up of the Mandelbrot set, a magnification of a tiny part of the original." He pointed to little black specks within the spirals. "Each of these bumpy-warty bits would look just like the original bushy shape if we magnify them. Currently we're looking at a section magnified 1,000 times. It's called the Valley of the Sea Horses because the spirals resemble sea horses' tails."

The spirals, with their organic nodes and turns, enchanted Herman. There were white and green spirals upon aquamarine spirals in an endless, intricate cascade. Here and there was a touch of crimson. The colors were so vivid as to nearly make Herman weep.

The computer was like a microscope opening a portal to a vast, unexplored, and unpredictable universe. Who would have thought such beauty could be hidden within what was initially so lumpy looking? How could such a simple formula produce an inexhaustible reservoir of magnificent shapes and forms?

Herman's face had always been sharp-featured and deeply furrowed, but right now when he caught a reflection of himself in the monitor, his face seemed more relaxed. It must have been the invigorating day he was having.

"Let's magnify this spot a bit more," Malcolm said, pressing a button on the keyboard. As Malcolm continued zooming in, one of the spirals in the sea horses' tails seemed to unfurl and blossom, exploding outward to angular infinity. It reminded Herman of ocean waves. He could almost smell the surf. The space was so vast! *Freedom* was the only word that came to Herman's mind.

For a second, Herman thought he saw a sketchy image of a young woman lounging on one of the fractal arms in the Valley of the Sea Horses. Her long fingers trailed lingeringly over one of the spirals. *Preposterous*, Herman thought. When he blinked, she was gone. The only thing that remained were her hypnotic, almond-shaped eyes—not on the screen before him but on the mental screen within his head. Herman looked around, but noone else seemed to have seen the beautiful woman.

Other people had come up, pulled by the beauty of the Valley of the Sea Horses. Herman was lost in the intricacy of his thoughts and didn't notice them, as Malcolm moved the magnification up to 5,000 times, focusing on another Mandelbrot lump. Herman regretted the passing of the valley, but he listened to Malcolm again. One of the few constant pleasures in Herman's life was learning about something new, something mysterious.

"Why, yes," Malcolm said, "fractals can exist in three dimensions, or more correctly, solids can have fractal dimensions. Let me show you a view of a fractal solid." He touched some keys and apologized for the slowness of his machine. Using just a 386, he was at the bare minimum needed for recreational mathematics. While he waited for the display, he said that fractals were a way of thinking, once you had seen them.

"What do you mean?" asked Herman.

"Well, fractals are everywhere—in the branching of the blood vessels that feed your heart, in the shaping of a snowflake, in Russian dolls within dolls. Once you see them, you see them everywhere: tiny structures within larger ones within still larger ones in a self-similar cascade."

Herman's eyes opened wider. "Can they exist across time?"

"You mean like patterns in the world? Stock market rises and falls? Sunspots, electrical patterns of the heart? I don't know. I've heard rumors, but never thought much about it."

Herman looked at the messy table beside the young man, and his eye glimpsed a woman's photograph on the jacket of a book.

"Who—who is that?" Herman said. The woman in the photo had beautiful almond eyes and a smile so radiant that it put a lump in Herman's throat. Her hair cascaded like curly fractals about her shoulders.

The young man grabbed the book, which was very dog-eared and full of bookmarks. "That's Kirsten Munchower. A fractal prophet of sorts. Wrote many books on fractals. Author of *Computers, Pattern, Chaos, and Beauty*, a best-seller in the technical world. She was a mathematician, mystic, and artist."

"Was?"

"Sadly, she died in her own computer lab," Malcom said, looking down. "Heart attack, the doctors said. She was so young. She was my age. Makes me feel inadequate, I've done so little. If we only knew more about life."

Herman's dream began then, and he dreamed and watched the fractal demonstration, and watched the demonstration and dreamed, until it was time to go home. Malcolm, who was glad to have sparked interest in another, gave Herman his business card. As Herman turned to walk away, he noticed that the disembodied robot hand was threatening to knock over the soiled "Fractal Tours" sign. He watched as Malcolm gently picked up the appendage and sent it scooting away in another direction.

It was three months before Herman called Malcolm. Herman was inquiring about the best over-the-counter system to display fractals, and wondered whether Malcolm would work for him as a paid tour guide.

Malcolm was a poor but dedicated game designer and writer. He had worked hard to earn the free-lance life, and he wasn't about to go back to an office. Nevertheless, he told Herman what to buy and where, and even offered to be his Virgil in the Inferno of computer sales. They drove to a large store, and Malcolm realized then just how wealthy Herman was. It was vast, real money that trifled with Malcolm's ambitions.

Herman bought a computer system that was more than the round-the-world tour Malcolm had bought himself two years before. That kind of money made Malcolm silent and timid, and Herman, who believed timidity was somehow equated with genius, began to think Malcolm a genius. So Herman grew expansive and told Malcolm his dream.

"My life has been very unhappy," Herman said. "I know it should have been happy. I was born rich. I went to the best business schools. My first wife was even richer than I, and she died, leaving me everything. I'm very good at making money, although I've never enjoyed it. I keep doing the same things: making money, making joyless marriages, trying hobbies, spending time with other 'winners' like myself."

"Iterations, like fractals," Malcolm whispered.

For a moment Herman didn't respond, and then he nodded emphatically. "Exactly. Until I heard your talk I'd been trying to figure out my life. Was it a sine wave of ups and downs? Or was it just a squiggle? Then I realized it was a fractal."

Malcolm's left eyebrow raised. "I had figured that out a long time ago," he said. "That's why I like some of the patterns. They quiet my soul when I gaze at them. I figure mankind developed mandalas because we couldn't do fractals."

"No," said Herman, "I don't want a meditation aid. I can buy meditation aids. I want an operative way to change my life."

"I don't follow."

"Well, even though my life has been largely unhappy," Herman said, gazing off into space, "it has had—as is inevitable in all lives—moments of happiness. I want to find those, emphasize them, spend all my time there."

Malcolm scratched his head. "Predicting the future seems unlikely, sort of biorhythms for the nineties."

"I don't want to predict the future. I want to *pinpoint* the past."

Malcolm listened to the rest of what Herman had to say. It dovetailed with a long-standing dream of his, so Malcolm abandoned his freedom and began working for Herman night and day on a way to chart Herman's life.

From time to time Herman's thoughts returned to the ephemeral vision he had had of the young woman, the fractal beauty he had come to think of as Kirsten Munchower exploring the fractal world in the computer display. He remembered exactly how she had sat on one of the spiral arms, and he even tried to locate the precise region of geometric space in which he had found her. He tried magnifications as great as a googol, which would have made an atom larger than the universe if applied to a physical system. But he never saw her again. The space was too immense for such a search. He told noone of his vision and soon doubted he had seen her there at all.

When Herman Poole was 72 and Malcolm Caswell was 42, and purists were pointing out that the twenty-first century didn't begin until the *next* year, the charting mechanism had been perfected. The research years had been good for Herman. The dull eyes of seven years before had been replaced by sparks of enthusiasm. He was—to his great surprise—much more wealthy now than before. Some of the studies of quasi-chaotic systems had had unforeseen applications in real estate and finance. In fact, after he completed his quest to understand all his past moments of happiness (what he thought of more and more as the philosopher's stone—much sought but seldom found), he intended to turn over his money engine to financing certain universities. He had discovered the secrets of money, and there was no reason why they couldn't help others make a profit.

Just as the years had been lush for Herman, they were lean for Malcolm. Malcolm's goals were different from Herman's. Herman wanted to discover which emotions and thoughts in his life had come before moments of happiness. He wanted to study the intricate boundary between misery and happiness so he could bequeath to the world the secret of happiness. Malcolm, on the other hand, wanted to discover the secret of creativity. Malcolm had written two good short stories, painted a good painting, and come up with a great game. But he had a hundred mediocre, derivative, awful experiences as well. If only he could learn the boundary conditions . . . . If he could learn when to proceed with an idea—when it would be great—then his life wouldn't be wasted. But the work was hard. He didn't like commitment and he feared failure. He had avoided marriage, deep friendships, permanent jobs, and even projects longer than 90 days before he went to work for Herman. The fact that he was working on another man's idea threatened to overwhelm him with depression. To conquer the depression he would work harder.

They had begun to look more and more like Laurel and Hardy. Malcolm resembled an unhappy Laurel, perhaps even a starving artist. Herman looked like Hardy— fatter and happier, ever more sure that the real secrets of happiness were about to be found. The charting mechanism they had developed would collect real-time data on Herman's life for further analysis, and also allow them to delve into the mysteries of his past.

Herman suspiciously eyed the cable of wires that led from a cap to a computer unit that emitted faint beeping sounds. He walked closer to the cap. "You want me to put that on?"

Malcolm smiled. "You didn't come just to look at it, did you?" He motioned for Herman to sit in the chair. "We won't be able to study your life until we perfect all this and do more experiments."

Herman settled into a comfortable leather chair. He liked the smell of new leather. "I look like Frankenstein," he said, tapping the cap on his head. "What do you call this thing?"

"A SQUID."

"Doesn't much look like a squid."

"S-Q-U-I-D," Malcolm said, adjusting the wires. "Stands for superconducting quantum interference device. It should be sensitive to tiny magnetic fluctuations in your brain. We're going to hook it to an electronic signal processor that can focus on any selected brain frequency, so we'll be able to lift a few memories from your brain's electrical and magnetic traffic flow."

They tried the machine for several hours. Occasionally Malcolm had the SQUID function in "reverse" mode. Instead of merely monitoring the tiny magnetic fields, it would periodically give Herman's hippocampus a brief burst of highly focused waves to trigger memories. It was like momentarily rewinding a tape recorder and listening for a few seconds as it played a song from the past.

"I can't figure this out," Malcolm said. "It's true that the more pulses I give to your hippocampus, the older the memory I trigger. But it's not a simple relationship."

Herman grinned. Had he really seen the relationship before Malcolm? "Give me 5 pulses."

"Hmmm, you suspect something, don't you?" Malcolm said. "OK, here goes." He placed his finger on a computer keyboard, typed a 5, and saw Herman's eyes dilate.

"I'm walking down Fifth Avenue. I see an old man carrying a bassoon and looking quite sad. Maybe it's because a pigeon just crapped in his instrument. I'm on my way to my banker." Herman looked at Malcolm. "That happened four years ago. Now gimme 8."

Malcolm typed 8 on the computer keyboard, hesitated for a second, and then hit Enter. Nothing happened. He typed 8 again and pressed Enter. Again Herman's eyes dilated.

"I'm in bed with Sharon, my last wife. I smell her perfume, full of lemon grass and jasmine. I feel the cool satin sheets, but she's not interested in anything physical. She's yelling at me for leaving all my fractal books around the bedroom." Herman stopped himself. "I know the exact date because we filed for divorce shortly after that. It was five years ago." Herman rubbed his eyes and added, embarrassed, "It had nothing to do with these experiments or fractals. We would've broken up anyway. We didn't have much in common."

"You seem tired," Malcolm said. "Let's experiment some more tomorrow."

"No," Herman said. "I've got the pattern. You give me 1 pulse, and I have an experience from a year ago. You give me 2 pulses, and I experience something from two years ago. Three pulses makes me go back three years."

Malcolm interrupted. "Yeah, but then it gets complicated. To go back four years you need 5 pulses. For five years, you need 8 pulses."

Herman nearly ripped the cap off his head. "Don't you see it? You're supposed to be the mathematical genius! 1, 2, 3, 5, 8, 13, 21, 34 . . . It's a Fibonacci sequence."

Malcolm's mouth hung open. "Of course. Today you are the genius! After the first two, every number in the sequence equals the sum of the two previous numbers." He paused. "Let's see, if I give you 55 pulses, you should experience something in your life from nine years ago."

Herman's grin stretched from ear to ear. "I suppose if you gave me enough pulses," he said jokingly, "we could do some experiments in past-life regression."

"We've quite enough to work on with your present life," Malcolm said before he realized it was a joke. He tried a witty comeback: "My only worry is that someday you'll get stuck in a moment in your past and not be able to return to the present."

Herman's smile faded. "Is that a real worry?" With his right index finger, he nervously started tracing patterns in the cool leather of the chair. There were some bad things in his past, some places he wouldn't send his worst enemy.

"Nah, I doubt it. A bigger worry is that I'll type the wrong number on the keyboard, something too large." He paused. "But what's the worst that could happen? You'd just be momentarily regressed to an embryo."

Herman tried to laugh it off. "You had better put some limiting numbers in that program code of yours. This is my life you're playing with." He stopped and then smiled. "I wouldn't want you accidentally typing some negative numbers."

Although on this day they had a major breakthrough, the experiment was not as easy going as they thought. For one thing, the Fibonacci sequence allowed them only to approximately locate events to the nearest year. They had to find a way to localize events more precisely in time. The larger problem, however, was that the brain was so noisy that Herman and Malcolm only picked up brief incidents, memories, occasional events that they could understand and analyze. For all their dreams to reach fruition, the SQUID would have to somehow be perfected. They needed more data.

It was 2010 and old men regretted that we had not lived up to Arthur C. Clarke's prophecy. Herman Poole was 82 and Malcolm Caswell was 52. The memory device had been perfected. Herman was tired. Ten years ago he had felt that the secret was about to be uncovered. Now he was ready for the game to end. He had managed a ninth marriage to one of the scientists. He had divorced her when he realized that she had been motivated by pity rather than greed. He would have ended the quest, but had no other life to go to. All he had was momentum.

He no longer had the great fortunes of ten years ago. Most moneyed people had made out well in the Belize conflict of '05, but his money had been in the wrong industries. He had gambled that humankind had learned kindness, that a certain gentility had developed. But as usual the forces of cruelty and stupidity had won out.

Malcolm had changed necessity into a sort of desperate virtue. He was close to abandoning all hope of finding the secret of creativity for himself, but with the energy that comes to men like a second wind in their fifth decade, he hoped the work that made him ragged might bring about a transpersonal good. Herman and Malcolm began to scan Herman's past.

They had found that they could more easily catalog Herman's memories and distinguish real events from the rest of the noise in his brain by first having Herman produce mu waves between 8 and 13 Hz. (This took Herman a few days of practice, but he soon learned to generate mu waves whenever he wanted to.) These brain waves arose at a few frequencies within the alpha waves produced by a wakeful but relaxed brain. Malcolm told Herman that mu waves were the resting rhythms generated by neurons in the sensorimotor cortex, a diffuse region of the brain that lay on top of the head, between the ears.

They used a neural network running on IBM's latest soda-can-sized supercomputer that learned to recognize the different brain patterns that occurred prior to each of Herman's memories. Thus they could predict, most of the time, whether a memory would be happy or sad before Herman actually had the recollection! They were now able to localize events in time with greater precision using their joint discovery of fractal Fibonacci numbers, which allowed them to probe between the integers.

Still, dissecting the memories and emotions, and cataloging the intricate patterns, wasn't easy. The information didn't seem to reside in any of the brain's well-charted magnetic and electrical pathways. Many memories seemed to be at frequencies higher than 30 Hz and were easily lost amid the rest of the brain's noisy electrochemical chatter.

Herman and Malcolm agreed on one thing: The technological forecasters of the past few decades had been correct. To imagine what computer technology would be like in the 2010s, the futurists had said, one had only to assume infinite computational power for free. It turned out to be true and, as a result, Herman's computers were powerful enough to isolate most of the important brain signals from the noise. The SQUID cap was now reduced to a narrow, fabric headband holding three postage-stamp-size receivers that could instantly triangulate on any relevant portion of Herman's brain.

Because computer hardware was virtually free, and computer memory and computational power virtually infinite, computer software was now the bottleneck in their operations. Luckily, reduced instruction set computing (RISC) had been replaced by hierarchical instruction set computing (HISC). The computing language C++++, now the rage, enabled Malcolm to easily make use of other program modules on other computers through the airwaves. But even with all this, even though they seemed so close to their dream, insufficiently complex software still held them back. Although they could plot all the changes in Herman's objective universe, they couldn't derive an equation or develop a fully functional neural net that could yield and interpret all the patterns.

Malcolm scratched his head. "If only we could get our hands on the latest FISC programs, we might have the needed complexity . . ."

"FISC?" Despite his age, Herman's dark eyes had the same penetrating light they always had. In his right hand he held a steaming cup of good Jamaican coffee to jump-start his heart.

"Fractal instruction set computers," Malcolm said wistfully. "With a single command, I could explore levels of complexity I can only dream of now."

"If you need more money—"

"Money's not the problem," Malcolm said. "It's just not available to buy yet."

"I think we're missing something very basic," Herman said in a scratchy voice. "We've been focusing on brain states. Maybe we've got to focus on—" He paused. "On something like spirit."

"Getting religious in your old age?" Malcolm kidded.

Herman gazed into the wisps and eddies of steam coming from his coffee cup. "How about hypnotic regression while using the SQUID?" he asked.

"Let's wait a few months. The new FISC operating systems should help."

Herman Poole didn't celebrate his 100th birthday because he had entered a coma the week before. One of the plasmid diseases that would cause future historians to label the 2030s the "Plague Decade" had done him in.

Malcolm Caswell was due to receive yet another honorary doctorate on that day, but chose instead to spend it with his friend of nearly 40 years. The University of Belize was giving 54 honorary doctorates for Statehood Day. Malcolm didn't know that the award praised him for his "endless creativity in finding new ways to model neurophysiological states," a phrase that biographers would unfailingly come to repeat. Instead, he watched his friend dying. The physicians told him Herman would die within a day.

Malcolm shivered as he gazed at the plastic tubes in Herman's natural and humanmade orifices. The tubes sustained Herman's life like a parachute slowing the descent of a falling body. The tubes delayed the descent into oblivion, but could not stop it. Herman dozed in and out of near consciousness as condensation collected on tubes in his nose.

"Any chance he can recover?" Malcolm asked Dr. Yamasoto, as he shifted his gaze nervously from Herman to the tubes to the doctor. Malcolm was wearing a hospital gown to protect Herman from any germs he carried on his clothing. On Malcolm's head was a plastic hospital cap.

"Unlikely," Dr. Yamasoto said. "All the signs suggest brain injury." Herman's pupils were slightly dilated and did not constrict when Yamasoto shined a light into them. He had no reflexes. "When we tried to take him off the respirator," Dr. Yamasoto said, "his body made no attempt to breathe on its own."

"What does that mean?" Malcolm asked. The antiseptic, phenolic odor of the hospital made him wrinkle his nose.

"The contraction of the diaphragm is a primitive brain function orchestrated by cranial nerves 3, 4, and 5. The fact that he couldn't breathe on his own suggests extensive neurological problems."

Could I have caused any brain damage with all our experiments? Malcolm wondered. No, Herman was old. He had to die one day. We all have to die.

An EKG machine in the corner of the room started to show chaotic electrical activity in Herman's heart. Malcolm tensed as he gazed at an amber bottle of fluid hanging from a pole by Herman's bed. It fed an IV line in Herman's right arm.

Malcolm stayed for another half hour, and sthere were a few moments of hope. Herman's left eye occasionally moved from side to side, although he did not appear to be aware of his surroundings. His diaphragm started rhythmic contractions, so he was removed from the respirator. But his favorable progress did not continue. He drifted in a twilight world, straddling life and death like a tightrope walker.

Malcolm gazed at Herman's blood, more brown than red, flowing through a flexible exsanguination tube and into a humming bypass machine. For some reason, the sound reminded Malcolm of the chanting of monks.

"Hello," an Asian nurse said to Malcolm. She turned on an array of halogen bulbs on the ceiling to help her see more clearly. She leaned over Herman's face, applying a lubricant to his eyelids.

"What's that for?" Malcolm asked.

"It prevents the eyelids from sticking together."

"Why's that a problem?"

"Comatose patients don't blink. They also secrete fewer tears, even when their eyes are closed."

A wave of sadness passed over Malcolm like a dark swell of ocean water. Dr. Yamasoto returned and filled a syringe with a cocktail of free radical scavengers and lazeroids and then injected the solution into a port in Herman's intravenous line.

Blood, heated to 99 degrees, moved with phenyl tertiary butyl nitrone through the IV lines and into Herman's body through a vein in his arm. The nurse dimmed the lights. Before Malcolm's eyes adjusted to the reduced illumination, all he could see were the green and red blinking lights on the cardiac monitors. Herman Poole, multimillionaire, was almost invisible amidst the Christmaslike twinkling.

"Dr. Yamasoto," Malcolm said, "I think I know something that might make his passage easier." The doctor listened to and agreed with Malcolm's proposal. It was simple enough to feed in the data. Malcolm wished he had thought of it earlier.

Malcolm Caswell fed Herman Poole his happiest moment. A few months before, while doing a series of zooms on the complicated manifold that represented Herman's subjective universe, they had found the brightest and most beautiful point.

"I think he's reacting to it," said Dr. Yamasoto. "But without your monitor I don't know how he's reacting."

"He seems to be reacting well," said Malcolm. "Look at his smile." They both gazed at Herman's happy face. Both his eyes were dilating.

"What's he experiencing?" asked Dr. Yamasoto.

"I don't know." Malcolm gazed into the red and green twinkling of the hospital monitors. "We know it's a happy moment, but we hadn't characterized it."

Herman was pulling up a dented, rusty folding chair at Fractal Tours. A young black man was demonstrating a particularly beautiful section of the Mandelbrot set known as the Valley of the Sea Horses. With its strange, knotty spirals it was the most beautiful thing Herman had ever seen. It brought him a sudden sense of wholeness.

"He is dying now," said Dr. Yamasoto.

Herman leaned forward, and the screen dissolved, and he flowed into the beauty and the strangeness. Swirls upon swirls upon swirls. The lovely pounding of the surf was the only sound. He smiled as a pleasant ozonic smell of low tide filled his nostrils. There is a fractal heaven, he thought. This is my new home.

From the corner of his eye, Herman saw the woman with curly, fractal hair and beautiful, almond-shaped eyes. As soon as she turned her little boat toward him, Herman felt a giddy excitement. It must be ecstasy to be riding on this cool, vast ocean in such a small craft! A slender spiral trailed its fractal tendril over her smooth skin, and she laughed.

"He's gone," said Dr. Yamasoto. "It's sad you never found what you were looking for." Malcolm slowly nodded.

The infinite spiral waves split light into a million colored sparkles as the woman guided her craft toward Herman and waved. Herman thought her smile was radiant. He waved back and then looked above her head to the twinkling specks of black.

"Freedom," he whispered, and then he smiled as fiery orange light reflected off spirals on tessellated water, challenging the azure of an endless fractal sky. Herman wanted to say something to her, something profound, something that adequately expressed his emotions. His heart seemed to beat faster, and suddenly he turned to her, and gently held her hand.

"I know," she whispered, giving his hand a squeeze. Herman thought she looked like a goddess in the glorious explosion of rosy light. Her hand felt warm, and he could almost feel a cool, tender breeze. "Let's go," she said, tenderly guiding him into her craft.

Together they gazed into the shining and limitless heaven. The craft turned and pointed toward a symmetrical collection of intertwined aquamarine spirals. They were moving towards open sea.

Tears in his eyes, Malcolm began removing the equipment attached to Herman's body. There was a hush in the hospital room. *Time to go*, Malcolm thought. Turning, he walked down the hallway, his eyes sandy, his bones aching. His feet were still clad in elasticized hospital slippers, and they made a scratching sound, a sound almost like the surf in a faraway sea.

C H A P T E R

7

The digits of π beyond the first few decimal places are of no practical or scientific value. Four decimal places are sufficient for the design of the finest engines; ten decimal places are sufficient to obtain the circumference of the earth within a fraction of an inch if the earth were a smooth sphere.

— Petr Beckmann, *A History of Pi*

The Million-Dollar, Trillion-Digit, Pi Sequencing Initiative

Ants, God, and Pi

Whenever I hear someone mention π—the ratio of the length of the circumference of a circle to its diameter, and the most famous mathematical constant of all time—I think of ants. The reason for this will become clear as you read further.

As background, the digits of the transcendental number[1] never end, and no one has detected an orderly pattern in their arrangement. Here are a thousand digits of π that I computed using the computer program at the end of this chapter:

pi=3.14159265358979323846264338327950288419716939937510582097494459230781640628620899862803482534211706798214808651328230664709384460955058223172535940812848111745028410270193852110555964462294895493038196442881097566593344612847564823378678316527120190914564856692346034861045432664821339360726024914127372458700660631558817488152092096282925409171536436789259036001133053054882046652138414695194151160943305727036575959195309218611738193261179310511854807446237996274956735188575272489122793818301194912983367336244065664308602139494639522473719070217986094370277053921717629317675238467481846766940513200056812714526356082778577134275778960917363717872146844090122495343014654958537105079227968925892354201995611212902196086403441815981362977477130996051870721134999999837297804995105973173281609631859502445945534690830264252230825334468503526193118817101000313783875288658753320838142061717766914730359825349042875546873115956286388235378759375195778185778053217122680661300192787661119590921642019 9 . . .

If the digits were random,[2] we would expect to find that, among other things, each digit occurs with equal frequency. And this is precisely what we find. For example, there seem to be about the same number of 3's as there are 1's, 2's as 4's, 0's as 9's, and so on.

The quest to know π in as much detail as possible seems almost spiritual. Humans, through the ages, have passionately attempted to calculate more and more digits. In 1579, Viete computed the value of π to a remarkable ten figures, and today humans know the value of π to more than 1 billion digits, thanks to modern computers.

When I see thousands of digits of π spread out before me on a computer printout, I defocus my eyes for a few seconds and imagine little ants swarming over the pages, each ant with a little painted digit of π on its abdomen. The ants begin to line up in rows along the page as they march in mindless synchrony. If ant number 100, carrying the 100th digit of π, were to switch positions with ant 101, who would know? Who would care? As I mentioned, the digits of π appear random, and certainly no calculations we humans perform using π would be affected by such a minor change. If God were a geometer, would he or she care whether some of the digits of π were switched? Would God return ant 101 and ant 100 to their original positions and set them marching again in perfect order, or would God just chuckle and let them parade along, happily, but out of order? Do the digits of π show no more pattern than would be produced by a gambling God playing a ten-digit roulette wheel for a long time (perhaps before our universe was created in the Big Bang)?

I feel more comfortable about the enormous number of digits of π by thinking of a surreal ant colony that carries all digits of π on the ants' backs. Each ant carries a single digit. Again, think of the ants lined up in rows on a page. The first ant has the first digit of π, a 3, painted on its back. The next has a 1; the next, a 4; and so on. We want all the ants to fit on a single page, so some ants will have to be smaller than others. Let's assume that the first ant with a 3 is very large—1 inch in length. If the body of each ant is half as long as its predecessor, the complete value of π with *all* its digits fits on a single page of ants, but not even the most powerful electron microscope will reveal the last digit. (The reason all the ants can fit on a page is because the infinite sum 1/2 inch + 1/4 inch + 1/8 inch + . . . is equal to 1 inch.)

The ants can compress the infinite digits of π in another way. Let us imagine that the ants can speak by manipulating their crude jaws. The first ant in the long parade of ants screams out the first digit, "3." The next yells the number on its back, "1." The next yells "4," and so on. Further imagine that each ant speaks its digit in only half the time taken by the preceding ant. Each ant has a turn to speak. Only the most recent digit is spoken at any instant. If the first digit of π requires 30 seconds to speak (due to the ant's cumbersome jaws and little brain), the entire ant colony will speak *all* the digits of π in 1 minute! (Again, this is because the infinite sum 1/2 minute + 1/4 minute + 1/8 minute + . . . is equal to 1 minute.)

Astoundingly, at the end of the minute, there will be a quick-talking ant that will actually say the "last" digit of π! The geometer God, upon hearing this last digit, may cry, "That's impossible, because π has no last digit!"

The Grand Pi-Sequencing Project

Pi is not the solution to any equation built from a less than infinite series of whole numbers. If equations are trains threading the landscape of numbers, then no train stops at pi. — Richard Preston, 1992

My fascination with π stimulated me to conduct the following survey of scientists, mathematicians, students, and even science-fiction writers. Before reading their answers, how would you answer the following?

1. Would you be in favor of a project to calculate 1 trillion digits of π?
2. Would you like the National Science Foundation (NSF) to fund such a project at 1 million dollars?

3. If, as a result of this study, the distribution of digits in π were found to be non-random (that is, clumpy, nonuniform), would you consider this to be a profound finding?

Interestingly, such a task is not beyond the realm of possibility. In 1989, the Chudnovsky brothers, two Columbia University mathematicians, computed more than 1 billion digits of π using a Cray 2 and an IBM 3090-VF computer. Their 1,011,196,691 digits would stretch 1,580 miles if printed using this typeface. Toward the end of the summer of 1991, the brothers had computed π to more than two billion digits.

A Rupture in Geometry

Most numbers are "transcendental." Very informally, this means that they lack a simple definition. Most numbers with names are not of this variety, so 2, √3, and 7/5 are not such numbers. Indeed, most people only know the names of one or two transcendentals: the best known is π, the second is e. So many numbers are transcendental that if all the numbers were put in a barrel, it is good as certainty that the first number pulled out would be so. (In such a lottery always bet on the transcendentals.)
— Peter Borwein, Science, 1994

Before we look at the results of the survey, I think many of you would enjoy additional background on π to fully appreciate the respondents' answers. First, ask yourself, what is the importance of knowing the decimal values of π? Obviously you do not need to know the value of π at all to draw a circle, but π does appear in many kinds of geometrical engineering calculations. Note that the ancient Greeks and Italians did just fine when the decimal value of π was known only to a few decimal places. They built perfectly stable, beautiful buildings; drew precise circles (Figure 7.1); and made many mathematical discoveries.

In one of his greatest books, *The Measurement of the Circle*, Greek mathematician Archimedes proved that the circumference of a circle is less than 3+1/2 and greater than 3+10/71 times its diameter. A refined value of π was obtained by the Chinese much earlier than in the West. For example, some scholars have suggested that at the time the Greeks knew π to three digits, the Chinese knew the value to two or three additional digits.

Normally we think of π simply as the ratio of the circumference of a circle to its diameter. So did pre-seventeenth-century humanity. However, in the seventeenth century, π was freed from the circle. Many curves were invented and studied (such as various arches, hypocycloids, witches . . .), and it was found that their areas could be expressed in terms of π. Just look in any geometry handbook today. Why does π seem to be associated with so many curves seemingly unrelated to the circle? Do the measurements of the lengths and areas of *all* simple curved shapes involve π?

The number π, having roamed far from its original association with the circle, finally ruptured the confines of geometry altogether. Today π relates to unaccount-

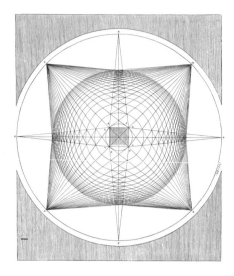

FIGURE 7.1. Circles within circles, by Renaissance architect Jan Vredeman de Vries.

ably many areas in number theory, probability,[3] complex numbers, and simple fractions, such as $\pi/4 = 1 - 1/3 + 1/5 - 1/7$ . . . . As another example of how far π has drifted from its simple geometrical interpretation, consider the book *Budget of Paradoxes*, where Augustus De Morgan explained an equation to an insurance salesman. The formula, which gave the chances that a particular group of people would be alive after a certain number of days, involved the number π. The insurance salesman interrupted and exclaimed, "My dear friend, that must be a delusion. What can a circle have to do with the number of people alive at the end of a given time?"

Three other π facts for you:

1. Thirty-nine decimal places of π suffice for computing the circumference of a circle girdling the known universe with an error no greater than the radius of a hydrogen atom!
2. The first six digits of π (314159) appear in order at least six times among the first 10 million decimal places of π.
3. In 1844, Johann Martin Zacharias Dase (1824–1861), a human computer, supposedly calculated π correctly to 200 places in less than two months:

π = 3.14159 26535 89793 23846 26433 83279 50288 41971 69399 37510
58209 74944 59230 78164 06286 20899 86280 34825 34211 70679 82148
08651 32823 06647 09384 46095 50582 23172 53594 08128 48111 74502
84102 70193 85211 05559 64462 29489 54930 38196

To compute π Johann Dase used: $\pi/4 = \arctan(1/2) + \arctan(1/5) + \arctan(1/8)$ . . . with a series expansion for each arctangent. Dase ran the arctangent job in his brain for almost two months.

Survey Results

We finally discuss the results of my survey. How do your own answers compare with the following? Do you think most respondents supported the project to sequence π?

I was startled to find that the majority of mathematicians and scientists were not in favor of a project to sequence the first trillion digits of π. Those against the project outnumbered those for the project in about a 4-to-1 ratio. In addition, a larger 5-to-1 majority was against the government funding the project at a million dollars.

As to the question of the importance of finding nonuniformity of the first trillion digits—for example, finding twice as many 0's as 1's—about 50 percent of the respondents thought this was profound. Some mathematicians from the Massachusetts Institute of Technology (MIT) said they would consider it "shocking" and profound if the first trillion digits of π were found to have a "non-normal" frequency distribution. (A normal number is defined as a number in which every possible block of digits is equally likely to occur; see previous endnote.) In other words, if there was a vast predominance of any one digit, or even a significant nonuniformity, they would consider it shocking. Many respondents suggested it was impossible to determine whether the digits were nonrandom because even in a trillion truly random digits, all sorts of patterns can *appear* to arise. I discuss this later.

Those in Support of the Sequencing Project

What follows are excerpts from some of the interesting comments I received from scientists, mathematicians, and science-fiction writers around the world regarding the π sequencing project. I thank them for permission to reprint excerpts of their comments here. First we examine the comments of those who supported such a grand initiative.

Many respondents indicated the project would be of value if it were done as an effort to research new techniques in computing, memory storage, and related computing issues. Chris Henrich wrote:

> As a computer engineer, I support the project. It would test several aspects of a computer system: correctness of software, reliability of hardware, reliability of storage media, and organization of a large-scale activity. One million dollars seems a small pricetag. If we keep within budget, we would compute 1 million digits per dollar. Like everyone else, I do not have any ideas about the distribution of digits in π. A million dollars spent upon getting such ideas would be a million dollars spent well.

Bill Beaumont, from the University of South Australia, wrote:

> It seems to me that a trillion digits of π would be a valuable resource well beyond their generation. Checking to see whether or not they were random would produce a host of new techniques in statistical tests for randomness, since the sequence being tested is much longer than any that has been available before. It would extend spectral analysis techniques, for example. Checking just an ordinary random number generator for randomness is already quite challenging. Exhaustively checking a trillion-digit sequence would be mind-boggling.

Don Webb, a science-fiction writer from Texas, is in favor of such an initiative for two reasons. He writes:

I'm more or less a neo-Platonist. First, I think that mathematical phenomena give insight into the deep structure of the world of thought as well as the material world. They provide a place of anagogic insight. Such a basic ratio as π provides a great deal of philosophical fuel. Second, I've yet to see any "pure" research that didn't pay off in some interesting tool for daily life. This mystery of achieving something powerful and useful while seeking some unmanifested intellectual concept is hard to explain but wonderful. Seek after the mysteries!

I asked Don Webb whether he would like the NSF to fund such a project at 1 million dollars. His response was:

Although as a taxpayer I'm sure we could get a better bid from the Chudnovsky brothers, I would say "yes." Of course it would be shot down in Congress. How do you sell a thing of romance to men of musty minds?

Moshe Gerstein would support the project for "national pride only, to prove that it can be done."

Dr. Gabriel Landini, an expert on the practical application of fractals and member of the Oral Pathology Unit in the School of Dentistry at the University of Birmingham, writes:

Yes, I would think that computing the first trillion digits of π would be an interesting task to conduct. I have the impression that this type of project is something that sooner or later is going to be done, but is it actually possible? A million dollars for such a project appears quite cheap. It would be 1×10^{-6} of a dollar per digit, or a million digits per dollar. However, there is a double problem: Is the problem worth studying? What else could be done with that money? I do not think that there will ever be a consensus on what research money is worth spending. What a research group considers "interesting" may be a waste of time and resources to somebody else. Many would argue that spending 1 million dollars on a project against malaria or global warming, or to improve education, or to fight crime would be much better spent than on "a series of numbers" for which a group of scientists has a particular interest. But what about the parallel activity that is generated in such a project? The "unexpected"? There is no doubt that solving problems like this promotes research into new algorithms or new computer architectures, and these may be applicable in other different and probably distant problems.

Yes, I believe in basic research. The other day, looking at a screen-saver on a colleague's computer display made me think of all those millions of CPU cycles lost on "nothing" (apart from the higher electricity bill). What about a "huge" parallel computation using the entire Internet? This would be similar to finding the area of the M-set using the Internet, as Jay Hill and Yuval Fisher are doing. What about a change of policy on the use of the Internet where you must "pay" for using the 'net by allowing others to share your own machine's unused computing cycles? This may help in the π project.

David Fowler, of the University of Nebraska–Lincoln, writes:

I would like someone to fund this. One million dollars is about a dime per person from New York City, not that most New Yorkers would give a dime to know the value of π to *any* number of digits. I'm concerned about the attitudes toward "curiosity-driven" research, for example, by Senate funding committees. To me, it is very important to maintain a certain level of curiosity-driven work. A π sequencing initiative might seem frivolous to many people outside mathematics and also to many mathematicians. The recent *New Yorker* article on the Chudnovsky brothers made clear the difficulty they've had getting funding. I think that we should have some exploratory missions like this. Certain fractal explorations should also be undertaken. Also, of course, we should sequence the number *e*.

David adds:

I think that our human role on this planet should be to try to understand the complexity of existence. What sort of explanations would we have without mathematics?

Those Against the Sequencing Project

As I mentioned, most respondents were against the sequencing project. Anton Sherwood, Libertarian candidate for California Assembly, District 12 (San Francisco), is against the public funding of such a project. He says, "If you want to do it without the taxpayers' help, I might kick in a dollar."

Professor Mike Frame from Union College was one of many mathematicians who did not support the project. He writes:

Given the current NSF budget outlook, I can't see why a million dollars should be invested to find a trillion digits of π (even though a million digits per dollar is a pretty good return). For most irrational numbers, I believe the digit sequence should be "random" (in terms of algorithmic complexity), but of course π isn't just any number. Could the geometrically natural form of its definition give rise to an eventual departure from randomness? Since nothing unexpected has been found in the several billions of computed digits, this suggests that if there is a connection between geometry and digits, it is subtle in human terms. Carl Sagan's *Contact* notwithstanding, I don't see what a departure from randomness in the digits of π would tell us. Since I'm a geometer and not a number theorist, you should apply an appropriate filter to my comments.

Although many mathematicians were against funding the project, some commented that they would support funding of the project more than they would support a project for studying the Mandelbrot set, a famous fractal object. Others said they did not favor the project but had no objection if the Chudnovsky brothers wanted to carry out such a task.

John Wilson writes:

I would find the discovery of patterns interesting, but there would be no guarantee that any patterns found wouldn't be broken after the trillion-and-first digit. I think the 1 million dollars could be better spent on things such as AIDS research, genetic engineering, or controlled fusion.

David Rusin, from Northern Illinois University, writes:

One million dollars would support at least half a dozen people like me over the summer, and I assure you we could produce much better math. I can't imagine anyone with sway in the NSF putting more than a sawbuck into this project. (Funding the development of a good machine *capable* of doing this might be of some interest.)

Daniel Winarski from Tucson, Arizona, did not agree that π is an important number. He believed that π was just the sum of the infinite series $\pi = 4 \times (1 - 1/3 + 1/5 - 1/7 + \ldots)$. There are, of course, an infinite number of other constants derived from other series. He says:

> Why is π important? Because the Greeks related it to the diameter of the circle millenia ago. But now it's time for a change.

Thomas Ward from Ohio State's Math Department writes:

> No, no, no, no. Apart from the terribly narrow interest of the project, the idea of million-dollar math projects in a country filled with bright Ph.D.'s who cannot get $30,000 post-docs is ridiculous.

Hugh LaMaster from NASA writes:

> I would rather see the 1 million dollars used to support mathematical researchers, rather than this kind of project. Now, don't get me wrong. Sometimes, "Big Science" projects are the only way to get support for researchers, and basic research, so if the 1 million dollars is actually going into mathematicians' salaries, then fine. It is unfortunate that it is much easier for politicians and the general public to understand big projects than basic research, and basic research is really suffering today. But, it seems to me that it would be very hard for the public to get excited about this project. Who needs more than five or six digits of π in their daily lives?

As for funding of the project, Hugh LaMaster writes:

> I would not be in favor of funding this project because I think it can be accomplished "for free" using spare CPU cycles on existing machines, and in any case the cost seems excessive to me, considering that the significance would be difficult to interpret.

Arlin Anderson from the Intergraph Corp. in Huntsville, Alabama, writes:

> It would cost you almost the entire million dollars just to *store* the answer! Since the computation *can* be done, I am sure someone will eventually do it. I support research, but I would rather hunt for an odd perfect number, or compute $1 + 2^{-3} + 3^{-3} + \ldots$ to about 50 digits and see if I could find a closed-form expression for the limit. (Note: $1 + 2^{-4} + 3^{-4} + \ldots$ has been done. It is $\pi^4/90$, which is why I find $1 + 2^{-3} + 3^{-3} + \ldots$ all the more intriguing.) Even though it may be difficult or impossible to demonstrate that π has random digits, we can show that a particular pattern is "statistically significant" if it satisfies certain mathematical criteria.

Simon Bradshaw from the United Kingdom writes:

> I would not like the NSF to fund such a project for 1 million dollars because, given the established trend in computer technology, such a project will cost $500,000 next year, and $1,000 in 2004. Unless there is a demonstrated need to know this information, its determination should wait until it is cheaper.

Randomness

Many respondents had difficulty with the question of randomness in the digits of π. In fact, many said the question was meaningless. Others said it would be possible to apply the same kinds of statistical tests to π as computer scientists and mathematicians apply to random number generators in an attempt to see whether π would make a good random number generator. Still, most agreed the interpretation of any findings would be very unclear.

Tim Chow from MIT notes that there is no way a finite number of digits of π can prove its "nonrandomness." All kinds of patterns can appear in a trillion numbers, but they do not prove that the numbers are nonrandom. However, if someone *proved* that the digits of π were "nonrandom" in some suitably precise sense, Chow thinks it would be "interesting and deep, though not necessarily important." He believes the usefulness of such a proof would depend on what mathematical techniques were used, and whether or not they generalized to other areas of mathematics and human endeavor.

Alon Drory also noted that we cannot determine whether an infinite-digit number is random. For example, he noted that random phenomena do produce clumpiness as the result of statistical fluctuations. Because π is infinite, any string of digits has a nonzero chance of being found in it. In fact, for example, it is certain that strings of 11111's of any finite length will appear in an infinite random-digit number. Alon writes:

> If the digits of π were found to be nonrandom, you might be able to describe them with a finite formula. The implications for mathematics would be certainly amazing, but I don't know about the world. (It would be equivalent to a proof that the axioms of the real numbers are inconsistent, since it would disprove that π is transcendental, which is a basic mathematical theorem.)

When I asked Alon Drory what he would say if π suddenly turned to all zeros after the trillionth digit, his response was:

> If π did run out *totally* after some digit, π would be a rational number. This has been mathematically proven not to be the case. If finding a trillion zeros means that *all* the remaining digits are also zero, this is logically equivalent to assuming that π is what it has been proven not to be. This in turn would mean that the real numbers are an inconsistent logical system. It is a theorem in logic that, in logically inconsistent systems, you can prove *anything* in this system. In particular, note that the algorithm used to actually calculate the digits of π relies on the theory of real numbers, so if this theory is logically inconsistent, there's no reason for one to believe that the algorithm means anything. We are locked in a logical contradiction.

George Schmitt of the Connection Machine Facility at the Naval Research Laboratory suggests that if the digits of π to some finite length appeared to have some sort of strange distribution, there might be something interesting behind it. He writes:

> Of course further study of the first 100 trillion digits of π would have to be done to determine what is really going on. I think this problem receives far too much attention. I suggest instead that the NSF fund the exploration of the other natural constant, *e*.

V. Guruprasad from IBM writes:

> If π ceases to be random after some trillionth digit, hardly any everyday physics would be directly affected. It might have a terrible impact on some QCD computation in quantum

physics, it might make it more chaotic, in more than one sense, but those things can't affect the world "today." It would take at least a couple of decades before there was any real effect on mortals.

Tom Hayes, a graduate student at the University of Chicago Mathematics Department, writes:

Even if π starts with statistically unusual digits, so what? While this might be an interesting fact, a proof would be much more interesting, and even that, although perhaps beautiful, would not be of much real use.

Matthew M. Conroy writes:

If the last "7" digit in π were found to occur around 10 billion digits into the sequence, and no more up to the next trillion, one cannot conclude much about the asymptotic density of 7's among the digits of π. This lack of 7's would be quite curious, and perhaps would lead people to try to prove that the density of 7's in the digits of π is not 1/10 (and hence π would be not normal), but, on its own, it would prove nothing. The evaluation of digits of π has no value whatsoever—from a mathematical point of view. How to do it has a lot of interest for the algorithm designer, computer scientist, etc.

Peter T. Wang from Cal Tech writes:

If π were not normal, it would be rather shocking, because there is a substantial body of evidence that hints that it is normal. But calculating a trillion digits doesn't accomplish much because it isn't a proof of anything other than the computer's ability to calculate. Nevertheless, a *proof* of the nonuniformity of digits would have a significant impact on mathematics, although hardly more than the actual proof of Fermat's last theorem. These two conjectures share one thing in common: The ideas and mathematical tools used to prove or disprove them is what is actually important, not their truth or falsehood.

Thomas Ward from Ohio State's Mathematics Department writes:

Would it really be a "finding" if we thought there was a pattern in π? That other fascinating transcendental number, Tom's number, $0.(10^{25}1s) + 10^{-25}\pi$, has wildly nonrandom behavior in its first trillion digits. I'm prepared to prove that for a mere $25. Please instruct NSF to send the cheque straight to me.

Mark Das from Berkeley, California, writes:

I think that finding true patterns in π would be important. Would it imply that the universe and all the numbers are not random at all, that e and $\sqrt{2}$ may not be irrational? I'm not sure what that means, but it would be really interesting.

Dr. Gabriel Landini (Oral Pathology Unit, School of Dentistry, University of Birmingham) writes:

If the digits prove to be nonrandom that would be very interesting. And of course, this would us allow to draw nicer, more accurate circles!

France Dacar from Computer Science Department of the Jozef Stefan Institute in Slovenia writes:

Even knowing a googolplex digits would not suffice, in itself, to decide whether the distribution of the digits of π is even or clumpy. One would need some theorem that would say something like: if the distribution of digits in π is uneven, then the first trillion digits have the following distinctive behavior (imagine here a description of some property). If the trillion digits fail to behave as in the theorem, then their distribution is uniform. Perhaps there would be a similar theorem that would help decide the issue in favor of clumpiness. But without even a whiff of such theoretical background there can be nothing more futile than computing all those digits.

Whatever opinion one has about it, nothing will deter a truly dedicated group of people from computing the digits. What's more, I believe that computing this trillion of digits is one of those great unavoidables, like death or the big crunch.

It would be quite another matter if somebody could decide whether the distribution of digits is uniform or not, by developing a succinct theory, preferably without using the computer. This could not help but give insight into lots of other things besides the number π and its digits.

Simon Bradshaw from the United Kingdom writes:

I would not find it important if the digits were found to be nonuniform. However, I know people who wouldn't consider the following to be important and profound:

1. The observed clumpiness of the cosmological mass distribution.
2. The hypothetical nonrandom content of genetic introns.
3. The hypothetical nonrandom decay of radioactive nuclei.

It's all a matter of your personal fields of knowledge and interest!

Don Webb, science-fiction writer, also wondered about the significance of finding patterns in π. He writes:

This is a tough question. This is the question of what is meaning and significance and pattern. There are patterns without meaning—the face of Elvis on Mars—but developing our own philosophical and mathematical tools to determine if nonrandom events are occurring would be a great achievement indeed. Of course if there were clear patterns, I would find it exciting and profound. I don't think I would ever look at a circle again without wondering what other patterns or meaning are hidden even in the most common things—not to mention the shape of the sun, the turn of my DNA, the vibration of superstrings, and the actuarial tables wherein human death is ruled by π.

Bill Beaumont from the University of South Australia thought that a discovery of strange patterns in the first trillion digits of π, or a proof that the digit arrangements had structure, would be profound and important. However, he believed that his colleagues would have different views on the proposed project. He writes:

Most of my colleagues in the education faculty here work from a critical theory perspective, which sees the purpose of research, not as finding out about the universe, but as redistributing power within society. To these people, the entire compass of science, mathematics, and technology is nothing but a positivist, masculinist plot to further oppress the disadvantaged. Although these people regard themselves as highly educated, most of them would not know what π was. They would not understand nor support the π sequencing initiative. Such people, sadly, have a stranglehold on our education system. My colleagues, of course, use electricity, drive cars, and do their wordprocessing on computers. They cannot see the irony in the fact that, to express and spread their condemnation of science, they need to use the products of science all the time.

Microbes, Hermann Schubert, and Pi

I close this chapter with a thought from more than a century ago. Hamburg mathematics professor Hermann Schubert, in 1889, described how there is no practical or scientific value to knowing π to more than a few decimal places:

> Conceive a sphere constructed with the earth at its center, and imagine its surface to pass through Sirius, which is 8.8 light years distant from the earth [8.8 years × 186,000 miles per hour]. Then imagine this enormous sphere to be so packed with microbes that in every cubic millimeter millions of millions of these dimunivitve animalcula are present. Now conceive these microbes to be unpacked and so distributed singly along a straight line that every two microbes are as far distant from each other as Sirius from us, 8.8 light years. Conceive the long line thus fixed by all the microbes as the diameter of a circle, and imagine its circumference to be calculated by multiplying its diameter by π to 100 decimal places. Then, in the case of a circle of this enormous magnitude even, the circumference so calculated would not vary from the real circumference by a millionth part of a millimeter. This example will suffice to show that the calculation of π to 100 or 500 decimal places is wholly useless.

Try It Yourself

Some of you may be interested in computing many digits of π. I could recommend implementing an infinite series, such as $\pi = 4 \times (1 - 1/3 + 1/5 - 1/7 + \ldots)$ in a computer program; however, standard computer languages would not give you the large precision needed to compute hundreds of digits. Personally, I recommend huge-precision languages such as REXX to do initial experiments. Included here is a REXX program that computes π to as many digits as desired. The program is inspired by the work of Samuel Lu.

C Program Code

```
/* REXX program to compute many digits of pi.
   To run, simply type the program name followed by
   the number of digits desired.              */
arg number_of_digits
numeric digits number_of_digits+2
interpret "error = 5E-"||(number_of_digits+1)%2
new = 1.414   /* Initial estimate for square root */
call sqrt 2
a = result; b = 0; pi = a + 2
do count = 1 until abs(pi_init-pi) < error
   pi_init = pi
   call sqrt a; y = result
   b  = y*(b + 1)/(b + a)
   a  = (y + 1/y)/2
   pi = pi*b*(a + 1)/(b + 1)
end
say 'pi='format(pi,,number_of_digits-1)
exit 0
```

```
/* Newton's method is used to find sqrt */
sqrt:
arg x .
do until abs(old-new) < error
    old = new
    new = 0.5 * (x/new + new)
end
return new
```

C H A P T E R 8

I believe that the infinity of possibilities predicted to arise in quantum physics is the same infinity as the number of universe-possibilities predicted to arise in relativistic physics when, at the beginning of time, the universe, our home, and all of its sisters and brothers were created. As modest and troublesome as we often are, we too are nevertheless creatures of infinity.

— Fred Wolf, *Parallel Universes*

If chess permits a virtually infinite variety of games, the rules of nature surely do. Science may be immortal after all.

— John Horgan

Infinite Chess

Chess is an ancient game for two players who move their pieces according to fixed rules across a checkerboard in an attempt to checkmate the opponent's king. Since its origin around the sixth century in India, the basic concepts have remained the same, although there have been occasional unusual variations. In my book *Mazes for the Mind: Computers and the Unexpected*, I discuss many bizarre and mind-boggling chess-game variants including relativistic chess, black-hole chess, circular chess, gun chess, ghost chess, fairy chess, madhouse chess, Martian chess, hexagonal chess, crushed chess, evolution chess, and carnivore chess.

One of my favorite chess variants is called "infinite chess," introduced to me by Tim Converse from the University of Chicago. In infinite chess, the game is played on an extremely large board or on one that extends arbitrarily in all directions. Queens, rooks, and bishops are permitted unlimited movement along their usual directions.

Imagine moving chess pieces on an "infinite" board, a section of which is displayed on a computer screen. This arbitrary-extent version of the game is well suited for play using a computer, because the computer can manipulate, magnify, and pan various regions of the board of interest. I am interested in hearing from any of you who have programmed such a playing board.

Tim Converse notes that this game would be very different from standard chess. For one, pawns would become extremely unimportant, and probably many of the initial moves would be long moves by the major pieces. When playing on a very large chessboard, pawns will not be converted to queens because it would be difficult for pawns to reach the last row. In the infinite version, pawns never reach a "last" row of the board.

If the chessboard were infinite—that is, played with no borders at all—the game could get very sparse as the pieces quickly disperse through the board. Infinite chess provides an interesting problem for game-playing computer programs, as the programs not only would have to limit the number of moves they look ahead (as usual), but would have to limit their "spatial" horizon as well.

Although infinite chess is fascinating to contemplate, my favorite alternate chess game is called Too Many Bishops, a game I invented in 1992. In this weird variant, each player acquires an additional bishop before each turn. The bishop is placed on a position determined by the opponent. The game ends when a king is captured, or when a player cannot move due to the mob of bishops on the board, or when a player cannot add a bishop at the start of a turn because all squares are occupied. (In an alternate version, if a bishop cannot be added because all the squares are occupied, the game continues as usual, until a bishop can be placed.) For clarification, here is how each turn starts: (1) Your new bishop is placed on the board at a position determined by your opponent; (2) next, you are free to move the new bishop or any of your other pieces as usual.

A somewhat related problem requires us to determine how many bishops can be placed on an extremely large, empty chessboard so that no piece can capture another. Before considering boards consisting of billions of squares, let's consider the more manageable case of an ordinary 8 by 8 board. It turns out that there are 14 ways in which bishops can be placed on an 8 by 8 board so that no piece attacks another. Here is one possible arrangement:

Convocation of Bishops (1)

```
B * * * * * * *
B * * * * * * B
B * * * * * * B
B * * * * * * B
B * * * * * * B
B * * * * * * B
B * * * * * * B
B * * * * * * *
```

Here is another one:

Convocation of Bishops (2)

```
* * * * B * * *
B * * * * * * B
B * * * * * * B
B * * * * * * *
* * * * * * * B
B * * * * * * B
B * * * * * * B
B * * B * * * B
```

Can you find any more bishop arrangements where no bishop attacks another? What if I were to ask you next to consider an incredibly large chessboard, say, 1 million by 1 million squares? Is it possible to compute the maximum number of bishops that can exist on such a board without attacking one another? What would such a convocation of bishops look like?

Joseph Madachy, in his book *Madachy's Mathematical Recreations* (1966), gives numerous interesting facts regarding this class of problems. He points out that the maximum number of bishops that can be placed on an n-by-n board so that none attacks another is $2n - 2$. The total number of ways of arranging these pieces is 2^n. Thus there are 256 ways of arranging the 8 bishops on a chessboard; however, due to rotations and reflections, only 36 of these solutions are distinct.

The number of distinct solutions for the bishops problem on an n-by-n board is

$$2^{(n-4)/2} \times (2^{(n-2)/2} + 1)$$

if n is even, or

$$2^{(n-3)/2} \times (2^{(n-3)/2} + 1)$$

if n is odd.

What is the ratio of the total number of solutions to the number of distinct solutions for infinite boards? The BASIC and C code in Appendix 1 allows you to compute this ratio and other values (Table 8.1) that show various solutions to the bishop problem. In Table 8.1, it appears that the ratio converges to a value of 8 for an infinite board.

I am interested in seeing sample bishop-convocation patterns from readers for boards larger than 8-by-8. Can any readers find a pattern for a 10-by-10 board? What is the largest pattern you can draw on paper?

TABLE 8.1.   Crowded Convocations of Bishops

n	Max. Number of Bishops	Total Number of Solutions	Number of Distinct Solutions	Ratio
2	2	4	1	4.00
3	4	8	2	4.00
4	6	16	3	5.33
5	8	32	6	5.33
6	10	64	10	6.40
7	12	128	20	6.40
8	14	256	36	7.11
9	16	512	72	7.11
10	18	1,024	136	7.52
11	20	2,048	272	7.52
12	22	4,096	528	7.75
13	24	8,192	1,056	7.75
14	26	16,384	2,080	7.87
15	28	32,768	4,160	7.87
16	30	65,536	8,256	7.93
17	32	131,072	16,512	7.93
18	34	262,144	32,896	7.96
19	36	524,288	65,792	7.96
20	38	1,048,576	131,328	7.98

Digressions

Many have become chess masters—no one has become the master of chess. — Tarrasch

Mathematicians today have determined that a checkers game has an incredible 10^{20} possible variations. In chess, some estimate the number to be 10^{44}. The ancient Chinese game of Go has a walloping 10^{120} positions.

Arlin Anderson from the Intergraph Corporation in Huntsville, Alabama, responded to my bishop challenge and found an infinite number of solutions of the following form:

```
*  *  ...  *  *
B  *  ...  *  B
B  *  ...  *  B
B  *  ...  *  B
.
.
.
B  *  ...  *  B
B  *  ...  *  B
B  *  ...  *  B
```

Here the "..." symbols indicate the board can be extended by an arbitrary number of squares, and still the bishops will not attack one another.

Recursive chess boards. Imagine chess games that are played on a fractal nesting of squares. Beginners play with only a few nested rectangles, while grand masters play with many. Computer programmers may design programs that allow the board to be magnified in different areas, which permits convenient playing at different size scales.

C H A P T E R 9

While we have discovered several groups of natives who have no conception whatsoever of religion, we have never come across a race (no matter how far it happened to be removed from the center of civilization) that was completely without some form of artistic expression.

— Hendrik Willem Van Loon, *The Arts*, 1937

We are told that the shortest line segment contains an infinity of points. Then even the shell of a walnut can embrace a spatial infinity as imponderable as intergalactic space.

— William Poundstone

It was all a pattern, as surely as a spiderweb is a pattern, but a pattern does not imply a purpose. Patterns exist everywhere, and purpose is at its safest when it is spontaneous and short-lived.

— Anne Rice, *The Witching Hour*

The Loom of Creation

Computer graphics art is everywhere, from television commercials, to beautiful ray-traced images appearing in computer graphics magazines, to processed images from the edges of the known universe. However, I find that insufficient attention is given to computer art that students and computer hobbyists can easily produce given the simplest of line-plotting programs. This brief chapter is an attempt to renew interest in the inexhaustible reservoir of magnificent shapes made possible with the lowly line.

Consider a race of spider-beings named Mygalomorphs who spend their days spinning webs upon circular frames. Status in their society is based on the beauty of their webs. To create the web patterns, the spiders string a straight piece of web from one point on the circle to another. Usually the patterns are dull and uninspiring, and therefore most spiders are relegated to lower social classes.

One day, a rather intelligent Mygalomorph let a straight web piece amble around the circumference of the circle, the front end going six times as fast the rear. In other words, every time the rear end of the straight web moved one space, the front end moved six. After a few moments' contemplation, the Mygalomorph realized that by the time the fast end had completed one trip around the circle, the slow end had traveled just one-sixth of the way around. His web grew ever more intricate as he continued weaving. His forelimbs moved back and forth with lightning speed.

When he stood back and gazed at his creation, it was not some complicated, meaningless pattern but rather a five-lobed object that mathematicians on Earth call a ranunculoid (Figure 9.1). Amidst the intricate beauty of the strands, a ghost of the ranunculoid seemed to materialize as if out of thin air! After many experiments, the Mygalomorph noticed that if one end of the web strand went n times faster around the circumference of a circle as its other end, then the web created a curve with $n - 1$ lobes. So beautiful were his patterns that the wise Mygalomorph soon became king of the spiders.

What strange shapes can you create by hand or by using the C or BASIC spider programs in Appendix 1? Both programs compute the endpoint positions (x, y) and (x_2, y_2) of each straight web chord on a circle. To control the programs, you can alter the values for variables a and b, which allows you to create an amazing array of spider forms. These two parameters control the speed with which one end of the

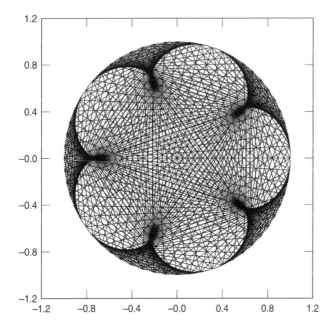

FIGURE 9.1.   Ranunculoid.

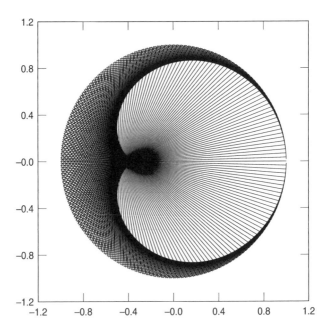

FIGURE 9.2.   Cardioid.

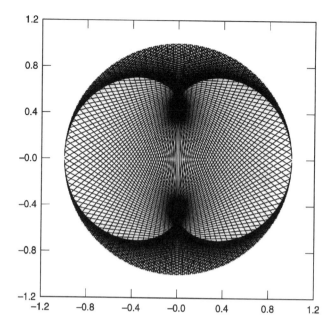

FIGURE 9.3. Nephroid.

spider strand moves with respect to the other. Figure 9.2, for example, was created by having one end of the web strand go twice as fast as the other ($a = 2$, $b = 2$). This produces a heart (cardioid) shape amidst an intricate weave of lines. Figure 9.3 was created using ($a = 3$, $b = 3$). It's called a nephroid. Figure 9.4 was created using ($a = 100$, $b = 100$), and, of course, this fabulously detailed shape has no common name. Figure 9.5 was created by using different values for a and b ($a = 2$, $b = 3$). Although it is quite beautiful, the points on this fishtail object do not lie on a circle. Can readers guess what constants were used to generate Figure 9.6?

Of all the web shapes in this chapter, the cardioid is the most famous. The cardioid (meaning heart-shaped) was first studied in 1674 by astronomer Ole Romer, who was seeking the best shape for gear teeth. When a circle rolls around another circle of the same size, any point on the moving circle traces out a cardioid. The Greeks used this fact when attempting to describe the motions of the planets. Finally, the cardioid is the envelope of all circles with centers on a fixed circle, passing through one point on the fixed circle.

What strange new worlds can you create using the spider programs? What happens when you use ever larger values for a and b? I am interested in hearing from readers who have discovered parameters that yield particularly beautiful and novel shapes. Two books also catalog interesting designs generated from simple algorithms.[1, 2]

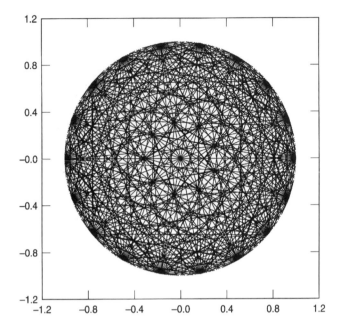

FIGURE 9.4.   Amazingoid.

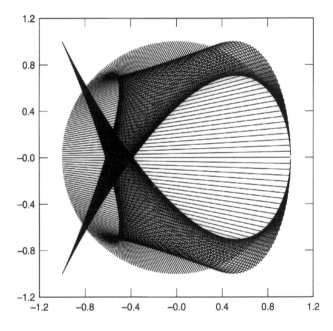

FIGURE 9.5.   Fishtailoid.

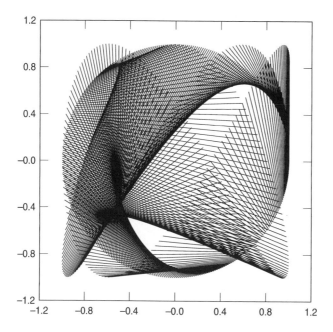

FIGURE 9.6. Enigmoid.

Slides in Hell

On a breezy April day in 1994, I was visiting a local playground, where I came across the most startling piece of recreational equipment. It was a metallic slide with a large hole punched in the chute. To this day, I cannot think of any reason it was there, because it seemed rather dangerous. A child usually slid from the top of the slide to the bottom, but it seemed as though a small child could get stuck in or even fall through the hole in the chute.

The image of this strange slide stayed in my mind for several months and stimulated the following mental exercise. Consider a metallic slide punched with ten large holes that are equally spaced from top to bottom. If we were to construct such a porous slide, and if the holes were large enough, you would certainly fall through at least one of the holes during your descent.

For this uncomfortable problem, let us assume that each time you encounter a hole you have a 50 percent chance of sliding through it and into a vile, oleaginous substance beneath the slide. For example, when your body passes above the first hole, 1 out of every 2 times you would expect to fall through the hole into the loathsome liquid below.

At first I asked colleagues the following question. You are offered 1 thousand dollars if you can correctly guess which hole a person would fall through as he or she descended the slide. Which hole would you choose? Soon the number of questions started escalating. Here's a summary of the most interesting ones I asked.

1. If you were a gambling person, which hole would you bet a person would fall through?
2. If you were a gambling person, how many attempts would you predict it would require for a person to slide from the top of the slide to the bottom without falling through a single hole?
3. If all the people on Earth lined up to slide down a longer, more horrifying slide with 100 holes, at a rate of 1 person per second, when would you expect the first

person to arrive at the bottom of the slide without falling through any hole? An hour? A day? A decade?

In Appendix 1, I include some exciting BASIC and C programs for simulating an interesting aspect of this process. First, let us attack the problem without the simulation.

It turns out that an analysis of this problem is not too difficult. The chance of falling through the first hole is 50 percent. The probability of reaching and falling through the second hole is lower. Specifically, it is the probability that a person did not fall through the first hole times the probability of falling through the second hole: $0.5 \times 0.5 = 25$ percent. For the third hole, we have $0.5^3 = 0.125$. This also means that, in a large enough sample, half the people will fall through the first hole, one-quarter through the second, one-eighth through the third, and so on. We can create a table (Table 10.1) showing the percent chance of reaching and falling into each hole as a person descends the slide.

Your chances of reaching and falling into the final hole are slim indeed—only about nine-hundredths of a percent. Therefore, most mathematicians I asked said that because they were limited to selecting one hole, they would place their money on hole 1. In other words, individuals would be most likely to fall through this hole. However, one respondent said he would not bet at all because the best chance is only a 50/50 chance.

As for the second question about how many attempts it would take to reach the bottom without falling through any hole, note that the chances of reaching each succeeding hole are the same as those of reaching and falling into the previous one. This means the chances of passing over all the holes are the same as those of reaching and falling into the last hole. Therefore, the probability is .0009765625; that is, 1,024 attempts would be required by a person, on average, to reach the bottom. Some people I queried suggested they would bet on a number less than this because the actual events can happen in any order.

TABLE 10.1.   Do You Dare?

Hole No.	Percent Chance of Reaching and Falling into Hole
1	50
2	25
3	12.5
4	6.25
5	3.125
6	1.5625
7	0.78125
8	0.390625
9	0.1953125
10	0.09765625

J. Theodore Schuerzinger from Dartmouth notes that because $1/2^{10}$ people would make it all the way down the slide without falling through any of the holes (1/1,024), this means that 1,023 out of 1,024 people would fall through a hole. He notes:

> Using the formula $(1{,}023/1{,}024)^x = 1/2$, we can determine that out of the first x people to go down the slide, there is a 50 percent chance that one person will make it down without falling through a hole. The solution $x = 709.4$ satisfies the equation. Thus I would bet that a person would make it all the way down on one of the first 710 attempts. In other words, after 710 attempts, the chance of someone succeeding exceeds 1/2.

What happens if all the people on Earth lined up to go down the 100-hole slide at a rate of 1 person per second? The probability of falling through the last hole is a minuscule $0.5^{100} = 7.88 \times 10^{-31}$. The average number of tries that a single person must make to finally arrive at the bottom is $1/(0.5^{100}) = 1.26 \times 10^{30}$, or, to be more precise, it would require 1,267,650,600,228,229,401,496,703,205,376 attempts, on average. (Your gluteus maximus muscles had better be in good shape before you attempt such an experiment.) In other words, we may see one person slide down over all the holes every 1.26×10^{30} attempts. At 1 person per second, we can compute that it would require, on average, about 4×10^{22} years, a period much longer than the age of the universe. Note also that this would require many more people than the world population can offer. Therefore, no one will reach the bottom of the slide, or, as one respondent said, "They will all try and fail (assuming everyone only gets to try once), or die waiting in line." Using a similar logic as for the ten-hole problem, after $(2^{100} - 1)/(2^{100})$ attempts, the chance of someone succeeding exceeds 1/2.

Simulation

The program codes in Appendix 1 allow you to mercilessly simulate the process of launching individuals down the slide. In particular, the programs create a distribution, shown in Figure 10.1, of the number of attempts it takes a person to slide down the entire slide without falling through a hole. The program simply continues to launch people down the slide and catalogs how many attempts it takes a person to succeed in reaching the bottom. The distribution is at first quite unexpected considering that, on average, 1,024 attempts are required to reach the bottom without falling through a hole. There is no peak at 1,024—rather there is a gradual "radioactive" (Poisson) decay from low values of the number of attempts to high values.

Steve Czarnecki from Owego, New York, was the first person to provide an interesting explanation for this mystery. To understand his argument, we define a Bernoulli trial as a random experiment with only two possible outcomes. The person sliding over a hole is a Bernoulli trial; the individual will either drop or pass over the hole with probability p and $(1 - p)$, respectively. Therefore, to make it all the way to the bottom, the person must achieve the "pass over" result of all the individual Bernoulli trials. This probability is given by $P = (1 - p)^{10}$, which is 1/1,024 when p is 1/2.

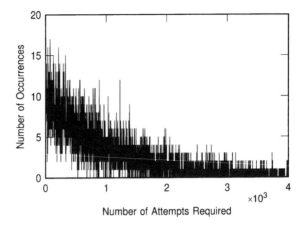

Number of Attempts Required

FIGURE 10.1. Distribution of the number of attempts it takes a person to slide down the entire 10-hole slide without falling through a hole.

Czarnecki next asks us to forget about the details of the slide itself and instead only observe the people climbing up the ladder to the slide, and also observe whether or not they appear at the bottom (after an appropriate time interval). This means we are observing *another* random experiment with two possible outcomes: Either the person makes it or does not make it to the bottom. In other words, every time a person tries the slide, it is a Bernoulli trial with outcomes "made it to the bottom" or "did not make it to the bottom."

We can now interpret the slide problem as an example of a sequence of independent, repeated Bernoulli trials with probability P of success. Here a success occurs when a person makes it to the bottom of the slide. If we let the random variable N represent the number of unsuccessful trials (attempts) required *before* the first success for that person (that is, success is achieved on trial $N+1$), then N has a geometric density defined by the equation $P(N) = P(1 - P)^N$. Here, $P(N)$ is the probability that N unsuccessful trials occur before the successful trial. This is the "exponential decay" law observed in Figure 10.1. Steve Czarnecki notes that the expected value of N (the average number of attempts required before the successful trial) will be $E(N) = (1 - P)/P$. For the example with 10 holes, 1,023 unsuccessful attempts are required, on average, before the successful attempt. Thus, our intuition that the person will achieve one success out of 1,024 attempts, on the average, is correct.

However, the geometric probability distribution has surprising implications. For example, Czarnecki points out that the most likely number of unsuccessful trials is zero—in other words, regardless of the odds, success on the first try is always the most probable event! This has the practical consequence that in any game of chance involving betting, one is better off "betting the ranch" on a single hand, rather than spreading one's stake over multiple hands.

Digressions

Here are some additional questions for you to ponder.

1. How would your answers differ to the three main questions asked in this chapter if, after falling through a hole, you had a 50 percent chance of entering a tube that shot you back up to the previous hole on the slide?
2. How would your answers differ to the three main questions if the probability for falling through the holes decreased by 5 percent for each hole? This might correspond to a slide where the holes got smaller and smaller as you slid down.
3. After reading about my slide puzzle, Joe McCauley from Colorado suggested studying the popular game Chutes and Ladders, also known as Snakes and Ladders. The game is played on a 10-by-10 checkerboard. The board has 20 to 25 chutes and ladders that transport pieces from one square to another. Perhaps computer simulations, such as those discussed in this chapter, can answer questions such as, What is the minimum number of moves required to reach the finish? Also, what is the mean number of moves required to reach the finish?

Alien Abduction Algebra

Infinity is where things happen that don't.
— S. Knight

Nobody has ever domesticated mankind. We are thus a wild species, as wild as the day we first went howling across the savanna. Perhaps the self-taming process of becoming a civilized species did not tame us to visitors, but only to ourselves . . . and then not very well, given our violent history.
— Whitley Strieber, *Communion*

I recently posed an unusual question to colleagues after watching one of the classic *Twilight Zone* television episodes. The episode concerned an alien's attempts to abscond with a group of humans to a spaceship in order to breed them for food. My own scenario and problem were as follows. An extraterrestrial, birdlike creature comes to Earth, captures one male human, and takes him to a large spaceship hovering in the Earth's upper atmosphere. The creature realizes that the male is unhappy without a companion, so the next year the creature abducts one female. Each succeeding year the creature duplicates his removals of the preceding two years, stealing the same number of humans, of the same sex, and in the same order. Thus, in the third year, the creature captures a male and then a female; in the fourth year, a female, a male, and then a female; and so on. The sequence goes as M, F, M F, F M F, and so on.

I posed two questions to colleagues. First, is it possible to determine the sex of the one-billionth human taken? Second, would the captured males be satisfied with the ratio of females to males existing on the spaceship when the one-billionth human is taken? (I asked colleagues to assume that the humans do not breed for the duration of this experiment.)

It turns out that the sex of the nth human is not too difficult to compute. In fact, in 1957, an obscure little paper was written on this class of problems, and a generating formula was discovered by T. F. Mulcrone of Loyola University. The Mulcrone

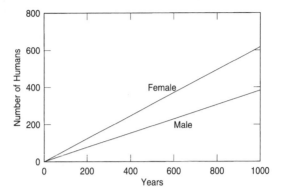

FIGURE 11.1.   The number of male and female humans changing through time.

formula can be adapted to my questions as follows. We denote a male by the number 1 and a female by the number 2. The sequence of males and females then becomes

$$M_n = 1,2,12,212, \ldots$$

The nth term M_n can be quickly computed from Mulcrone's formula:

$$M_n = [kn] - [k(n-1)],$$

where

$$k = (\sqrt{5} + 1)/2,$$

and the brackets indicate an integer truncation. In other words, $[x]$ is the greatest integer not exceeding x.

C and BASIC program listings are provided in Appendix 1 so you can compute the nth term of the sequence. You can also use these programs to compute the ratio of males to females by maintaining a count of their numbers through time. Figure 11.1 shows the growth in the number of males and females through the years, computed using the program codes.

Armed with these simple yet powerful programs we can now determine the sex of the one-billionth abductee. If the first abduction is considered to have taken place in year 0, then, using the Mulcrone formula, you can determine that it takes only 42 years for the alien to accumulate 1 billion humans. (This is about one-fifth of the world's population and roughly the number of Chinese.) The ratio of females to males is 1.618 to 1, which, as one computer programmer from Boca Raton said, "is better than the ratio in most bars."

If you want to use the computer programs to compute sexes for a large number of years, it is important to have a high-precision value for $\sqrt{5}$, and you might want to check the value that is used in your particular computer language. (You do not have to worry about this issue if you only want to compute the sex of the first few thousand abductees.) For example, many people who tried to use the Mulcrone formulation computed that a female would be the billionth person taken. This is because BASICA gives a value of 2.2360680103 for $\sqrt{5}$ on some machines, whereas the true value is 2.236067977 . . . .

Notice that the number of males and females, and the total number of humans, begins to follow the well-known Fibonacci sequence:

Year	0	1	2	3	4	5	6	7	8	...
Number of males	1	0	1	1	2	3	5	8	13	...
Number of females	0	1	1	2	3	5	8	13	21	...
Total	1	1	2	3	5	8	13	21	34	...

The Fibonacci sequence of numbers (1, 1, 2, 3, 5, 8, . . .), named after the wealthy Italian merchant Leonardo Fibonacci of Pisa, plays important roles in mathematics and nature. These numbers are such that, after the first two, every number in the sequence equals the sum of the two previous numbers, or $F_n = F_{n-1} + F_{n-2}$.

Note that you can also take advantage of the following mathematical relation well known to Fibonacci aficionados: The sum of elements F_1 through $F(n)$ is $F(n+2) - 1$. Using this relationship, it is possible to show that the number of people abducted during a particular year is simply F_{year} (in this case, the first abduction is considered to have taken place in year 1). The total number of people abducted including the current year is $F_{year+2} - 1$. As to questions about the sex ratio, it is possible to show that the ratio of the number of females to males converges to

$$F(n)/F(n-1) = \varphi$$

Here φ is known as the golden ratio and is equal to 1.61803 . . . . It appears in the most surprising places in nature, art, and mathematics. The symbol φ is the Greek letter phi, the first letter in the name Phidias, the classical Greek sculptor who used the golden ratio extensively in his work.

To avoid any numerical precision problems that may arise with the Mulcrone formulation, R. Biyani has suggested a formulation involving only integer calculations. In particular, we can use a recursive function that computes the sex, s, of the xth person in the yth year using a previously generated sequence of the number of persons taken in each year (the Fibonacci sequence). The recursive relationship is:

$$s(y, x) = s(y - 2, x), \text{ if } x \le F(y - 2)$$
$$= s(y - 1, x - F(y - 2)), \text{ if } x > F(y - 2)$$

where $F(y)$ is the number of persons taken in year y and $s(y, x)$ is the sex of xth person taken in year y.

There are, of course, numerous fascinating observations that can be made regarding the abduction sequence. In fact, R. Helmink of Raleigh, North Carolina, has complied a huge list of them. For example, the last person taken each year is a female. The first person taken alternates: male and then female. The number of females taken in any year equals the number of people taken the previous year. The number of people taken in any year equals one more than the cumulative number from two years prior. The cumulative number of females at the end of any year equals the cumulative number of people from the previous year. The cumulative number of males at the end of any year equals one more than the cumulative number of females at the end of the previous year and also equals one more than the cumulative number of people two years prior. The cumulative number of people at the end of any year is equal to the sum of the cumulative number of people from the two preceding years.

Digressions

Or would we find an even greater mystery, that the whole pantheon of our reality was somehow contained in the wobbling mind of that creature, who fell down to thank her raw new gods after a panther leaped at her throat, and by a miracle missed devouring us all.
— Whitley Strieber, Communion

Here are some additional questions for you to ponder.

1. How many years would the alien require to remove the entire population of the Earth (about 5.5 billion people)?
2. Some colleagues noticed that my problem results in a recurring embedded pattern, 13 characters long: M F M F F M F M F F M F F. Can you use this fact to determine the sex of the one-billionth person?
3. K. Ling of Endicott, New York, has noticed a peculiar structure for each year. Particularly, (Year N) = (Year $N-2$)(Year $N-3$)(Year $N-2$). For example,

$$(Year\ 8) = (Year\ 6)\ (Year\ 5)\ (Year\ 6)$$
$$(FMFMFFMFMFFMFFMFMFFMF) =$$
$$(FMFMFFMF)(MFFMF)(FMFMFFMF)$$

Can you use this fact to write a program to compute the sex of the billionth individual without resorting to the Mulcrone formulation?
4. How do the sex ratios change if, during the first year, you start with *two* people, such as, M M or F M?

Abduction of Earth's inhabitants by a Mygalomorph. (Rendition by William Rowe.)

The Leviathan Number

None is so fierce that dare stir Leviathan up.
— Job 41:10

In my book *Chaos in Wonderland*, an extraterrestrial philosopher presents the following number to his disciples and asks them what is special about it:

182,687,704,666,362,864,775,460,604,089,535,377,456,991,567,872

After much discussion, one young insectile student speaks: "It is the first power of two that exhibits three consecutive 6's."

The audience of philosophers applauds with delight. In fact the large number with 666 in its digits is equal to 2^{157}. I have called numbers of the form 2^i that contain the digits 666 "apocalyptic powers" because of the prominent role 666 plays in the last book of the Christian New Testament. In this book, called the Revelation (or Apocalypse) of John, 666 is designated by John as the Number of the Beast, the Antichrist—and various mystics have devoted much energy to deciphering the meaning of 666.[1]

My fascination with apocalyptic numbers started in my book *Mazes for the Mind*, in which I began a computer search for apocalypse numbers. These are Fibonacci numbers with precisely 666 digits. As you may know, the sequence of numbers 1, 1, 2, 3, 5, 8, . . . , is called the Fibonacci sequence, and it plays important roles in mathematics and nature. These numbers are such that, after the first two, every number in the sequence equals the sum of the two previous numbers, or $F_n = F_{n-1} + F_{n-2}$. It turns out that the 3,184th Fibonacci number is apocalyptic, having 666 digits. For those of you who are numerologists, the apocalyptic number with all its digits is:

116,724,374,081,495,541,233,435,764,579,214,184,068,974,717,443,439,437,
236,331,282,736,262,082,452,385,312,960,682,327,210,312,278,880,768,244,
979,876,073,455,971,975,198,631,224,699,392,309,001,139,062,569,109,651,
074,019,651,076,081,705,393,206,023,798,479,391,897,000,377,475,124,471,
344,025,467,950,768,706,990,550,322,971,334,370,940,093,654,442,411,815,
206,857,904,041,043,400,568,568,081,194,379,503,001,967,669,356,633,792,

97

347,218,656,896,136,583,990,327,918,167,352,721,163,581,650,359,577,686,
552,293,102,708,827,224,247,109,476,382,115,427,568,268,820,040,258,504,
986,113,408,773,333,220,873,616,459,116,726,497,198,698,915,791,355,883,
431,385,556,958,002,121,928,147,052,087,175,206,748,936,366,171,253,380,
422,058,802,655,291,403,358,145,619,514,604,279,465,357,644,672,902,811,
711,540,760,126,772,561,572,867,155,746,070,260,678,592,297,917,904,248,
853,892,358,861,771,163

Interest in these kinds of numbers peaked when I asked programmers and scientists to consider the monstrous "Leviathan number," a number so large as to make the number of electrons, protons, and neutrons in the universe (10^{79}) pale in comparison. (It also makes a googol, 10^{100} look kind of small).

The Leviathan number is defined as

$$(10^{666})!$$

where the exclamation point indicates a factorial. The number derives its name from a huge sea animal often symbolizing evil in Christian literature, including the Old Testament. I first came across the Leviathan while perusing Psalm 74, in which God battles the Leviathan, a dragon or serpent of some kind: "Thou breakest the heads of Leviathan in pieces, and gavest him to be meat to the people inhabiting the wilderness." In the Apocryphal Book of Enoch we read, "In that day shall be distributed for food two monsters; a feminine monster whose name is Leviathan, dwelling in the depths of the sea, above the springs of waters; and a male monster whose name is Behemoth . . . ." The Leviathan number is also intimidating from a mathematical standpoint due to its probable incalculability, as we will soon see.

Recently I asked colleagues a number of questions pertaining to the Leviathan number. For example:

1. What are the first six digits of the Leviathan number?
2. Could modern supercomputers compute the Leviathan, or will this be beyond the realm of humankind for the next century?
3. Even if we cannot compute the Leviathan, how many other characteristics of this number can we write down?

Michael Palmer from the United Kingdom was the first person to determine the first six digits of this incredibly large number. Interestingly, you do not have to compute all the digits of the Leviathan to determine just the first six. The reasoning is as follows.

Factorial functions can be approximated by a formula called Stirling's formula. It is named after James Stirling (1692–1770), a Scot who began his career in mathematics amid political and religious conflicts. He was friends with Isaac Newton, but devoted most of his life after 1735 to industrial management. His ingenious formula for approximating factorial values is

$$n! = \sqrt{2\pi} \times e^{-n} \times n^{n + 1/2}$$

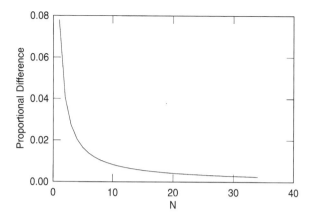

FIGURE 12.1. Percentage difference between the Stirling approximation and the actual factorial value as a function of n.

In Appendix 1, I provide BASIC and C code for computing Stirling approximations, actual factorial values, and the percentage difference between the two. Notice that this formula gives a useful approximation for $n!$ when n is large, such as when $n > 10$. For example, Table 12.1 lists values for the factorial function and the Stirling approximation so you can compare the two. The computations were carried out using ordinary floating point precision. Notice that the difference between the two actually increases as a function of n, but the percentage difference decreases with greater values. Figure 12.1 shows this *percentage* difference as a function of n. Interestingly, because many modern software packages allow us to compute large factorials (though presumably not so large as a googol), people often forget Stirling's formula. However, until a few years ago, this was the only way to approximately determine factorials for large numbers.

Let us return to how we can use Stirling's formula to compute the first few digits of the Leviathan without computing all the digits. Michael Palmer notes that for $n = 10^{666}$, the term $n^{n+1/2}$ in Stirling's formula is a power of ten and can be ignored when trying to determine the first six digits of the Leviathan. Next, let us look at the exponential term in Stirling's formula. Here we have $e^{-10^{666}}$, which can be rewritten as $10^{-10^{666} \times k}$ where $k = \log_{10} e$. Next, we split $10^{666 \times k}$ into its integer and fractional part, say m and f, giving us $e^{-10^{666}} = 10^{-m} \times 10^{-f}$.

We can ignore the 10^{-m} part because it is a power of 0.1, and therefore the first six digits of $10^{666}!$ are given by the first six digits of $\sqrt{2\pi} \times 10^{-f}$. Michael used a mathematical software package called AXIOM to compute this using a high number of digits (777) to ensure accuracy. Therefore, $10^{666 \times k}$ is 434294 . . . 9,652.27174945413317 . . . . Next, using what remains of Stirling's formula, we find $\sqrt{2\pi} \times 10^{-0.27174945} = 1.340727397$. Michael therefore concludes that the left six digits of the Leviathan number are 134072.

TABLE 12.1.   Stirling Approximations (S), Factorials (F), and Difference

N	F	S	Percent Difference
1	1	0.922	0.077863
2	2	1.919	0.040498
3	6	5.836	0.027298
4	24	23.5	0.020576
5	120	118	0.016507
6	720	710	0.013780
7	5,040	4,980	0.011826
8	40,320	39,902	0.010357
9	362,880	359,536	0.009213
10	3,628,800	3,598,695	0.008296
11	39,916,800	39,615,624	0.007545
12	479,001,600	475,687,488	0.006919
13	6,227,020,800	6,187,239,424	0.006389
14	87,178,289,152	86,660,997,120	0.005934
15	1,307,674,279,936	1,300,430,716,928	0.005539
16	20,922,788,478,976	20,814,114,062,336	0.005194
17	355,687,414,628,352	353,948,321,972,224	0.004889
18	6,402,373,530,419,200	6,372,804,291,198,976	0.004618
19	121,645,096,004,222,980	121,112,786,347,491,330	0.004376
20	2,432,902,023,163,674,600	2,422,786,791,066,042,400	0.004158
21	51,090,940,837,169,725,000	50,888,617,503,519,408,000	0.003960
22	1,124,000,724,806,013,000,000	1,119,751,437,820,100,600,000	0.003781
23	25,852,017,444,594,486,000,000	25,758,524,968,130,088,000,000	0.003616

Eternity

Could modern supercomputers compute the entire Leviathan, or will this be beyond the realm of humankind for the next century? Many have pointed out that the number of digits is more than 10^{668}, and this is much greater than the number of particles in the universe. Furthermore, even if a googol digits could be printed (or stored) per second, is would still require so much time that the universe would come to an end before the printing or storing were completed. Therefore, such a computation will always be beyond the realm of humanity.

Interestingly, David Radcliffe of the University of Wisconsin–Milwaukee Math Department has determined the number of trailing zeros of the Leviathan number. To solve this problem, it suffices to determine the number of factors of five among all the numbers between 1 to 10^{666}. For example, if a number is divisible by 5^k but not

5^{k+1} then it contributes k zeros to the product. Try this yourself for some smaller numbers.

The number of numbers between 1 and n that are divisible by d is $[n/d]$, where the brackets represent the greatest integer function. Using this, we can compute the number of zeros in $n!$ by the following incredible formula:

$$\sum_k = 1 [n/5^k]$$

If we eliminate the brackets, note that this formula is a geometric series whose sum is $n/4$. In fact, $n/4$ is a very good estimate for the number of trailing zeros. It can be shown that this approximation is off by at most $\log n/\log 5 + 2$.

Thus, David Radcliffe says, $10^{666}!$ ends in $2.5 \times 10^{665} - r$ zeros, where r is between 1 and 1,000. He used a software package called MAPLE (University of Waterloo), which supports high-precision arithmetic, to determine the exact value of r, and finds the number of trailing zeros to be $2.5 \times 10^{665} - 143$.

In other words, one can compute the exact value for

$$\sum_k = 1 [n/5^k]$$

in MAPLE by performing:

```
sum := 0:
N := 10^666:
for i from 1 by 1 while N>1 do
    N := trunc(N/5):
    sum := sum+N
od:
sum;
```

Carrying out this sum, we find that the number of zeros in the Leviathan number is:

24,999,999,999,999,999,999,999,999,999,999,999,999,999,999,
999,999,999,999,999,999,999,999,999,999,999,999,999,999,999,
999,999,999,999,999,999,999,999,999,999,999,999,999,999,999,
999,999,999,999,999,999,999,999,999,999,999,999,999,999,999,
999,999,999,999,999,999,999,999,999,999,999,999,999,999,999,
999,999,999,999,999,999,999,999,999,999,999,999,999,999,999,
999,999,999,999,999,999,999,999,999,999,999,999,999,999,999,
999,999,999,999,999,999,999,999,999,999,999,999,999,999,999,
999,999,999,999,999,999,999,999,999,999,999,999,999,999,999,
999,999,999,999,999,999,999,999,999,999,999,999,999,999,999,
999,999,999,999,999,999,999,999,999,999,999,999,999,999,999,
999,999,999,999,999,999,999,999,999,999,999,999,999,999,999,
999,999,999,999,999,857

which is just another way of writing $2.5 \times 10^{665} - 143$.

Superfactorials, Super-Leviathans, and Incomputability

Readers of this chapter will no doubt be interested in an additional class of large numbers called superfactorials. In 1987, A. Berezin discussed the mathematical and philosophical implications of the superfactorial function, defined by the symbol $, where

$$N\$ = N!^{N!^{N!^{\cdot^{\cdot^{\cdot}}}}}$$

(The term $N!$ is repeated $N!$ times.)

As we climb the integers in our quest for infinity, we eventually come to the super-Leviathan 9^{9^9}, which is the largest number that can be written using only three digits. It contains 369,693,100 digits. If typed on paper, it would require around 2,000 miles of paper strip. Since the early 1900s, scientists have tried to determine some of the digits of this number. Fred Gruenberger recently calculated the last 2,000 digits and the first 1,200 digits.

Even more unimaginable is the number $9^{9^{9^9}}$. If typed on paper, it would require $10^{369,693,094}$ miles of paper strip. Joseph Madachy has noted that if the ink used in printing the number was a one-atom thick layer, there would not be enough total matter in millions of our universes to print the number. Shockingly, the last 10 digits of this behemoth have been computed. They are 1,045,865,289.

Here are some other questions to ponder:

1. Is the apocalyptic Fibonacci number shown at the beginning of this chapter the only apocalyptic Fibonacci number?
2. Does there exist an apocalyptic prime number?
3. Is there any significance to the fact that the first four digits and the last four digits (1167 and 1163) of the apocalypse number at the beginning of this chapter are both dates during the reign of Frederick I of Germany, who intervened extensively in papal politics?

Leviathan fractal produced by methods similar to those discussed in chapters 6 and 23.

CHAPTER 13

Welcome to Worm World

*What if our science rests on irrational impulses that we cannot measure?
What if our mind is a ruler that cannot measure itself without always
getting the same answer?*

— George Zebrowski

Worm World is inhabited by large, serpent-like worms with forked tongues and coiling bodies. Imagine that this species of worm becomes seriously allergic to a substance the second time it is exposed to the offending material. From a biological standpoint, this scenario is not such a strange idea. In fact, it is similar to real allergic reactions in which the immune system becomes overly sensitive to an antigen upon re-exposure. (This sensitivity is sometimes known as anaphylaxis.)

In the hypothetical world described in this chapter, each worm wanders around while trying to avoid those allergens it has encountered previously—otherwise, it dies. You do not have to get dirty digging in your garden to study Worm World. Rather, the following simple exercises should have you wriggling with delight without ever leaving your home or office. Also, you do not need a computer to investigate Worm World, but BASIC and C code are provided in Appendix 1 for those of you interested in computer models. Although my simulations were all performed in a two-dimensional world, you can easily imagine experiments in three dimensions.

The game I designed to simulate the birth, death, and avoidance process is called the Anaphylactic Worm Game. I presented a quick example of this game as a puzzle in *Discover* magazine, but now would like to give you more insight and programming information. The game is really quite simple and is played on a chessboard filled with random numbers having values from 0 to 24. These 25 possible values represent 25 different allergens. Worms may start on any square on the board as they attempt to find the longest possible path through the board. The worms move horizontally or vertically (not diagonally). Each number along a worm-path must be different, or the worm dies.

Numerous fascinating questions arise in Worm World. For example, what is the longest path you can find in the three Worm Worlds in Table 13.1? After a few quick

TABLE 13.1.   Three Worm Worlds Used in the Internet Worm World Tournament

Worm World 1:

0	0	2	21	2	23	15	18
4	22	24	17	12	4	11	19
3	3	11	15	5	12	6	8
18	23	6	12	18	24	15	11
9	7	19	18	16	18	14	18
2	17	10	6	5	7	20	7
5	24	5	24	18	0	1	3
24	21	6	8	21	23	18	13

Worm World 2:

12	7	11	4	22	22	17	6
15	24	2	1	9	8	16	20
11	23	15	9	17	6	24	4
23	21	3	4	8	17	22	17
6	6	22	16	2	5	8	1
11	10	17	2	8	18	14	10
7	11	24	3	13	10	15	7
20	20	3	1	9	7	10	6

Worm World 3:

2	13	20	9	19	8	16	12
22	15	16	14	14	11	12	5
13	12	3	6	4	12	8	15
4	21	8	18	17	13	5	1
9	14	15	19	15	4	4	2
16	6	16	4	11	16	3	17
3	15	15	3	15	14	8	21
22	19	16	24	0	19	8	22

experiments using pencil and paper, you will see that worms die quickly in Worm world. But what is the quickest death you can find?

Before discussing the Worm World tournament I conducted on the computer networks, I would first like to tell you about simple computer programs you can write to explore the mysteries of Worm World. For example, the BASIC and C programs in Appendix 1 both start by filling a grid with random numbers between 0 and 25. (I used these programs to generate the three Worm Worlds in Table 13.1, but you can easily design worlds by hand.) Next, the programs attempt to find the longest possible path for each starting square in the 8-by-8 grid. To do this, they scan the im-

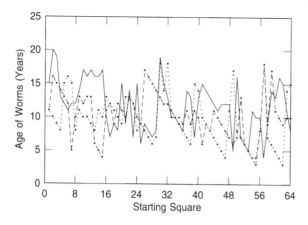

FIGURE 13.1.   Distribution of worm ages for all three Worm Worlds.

mediate neighborhood of a cell (up, down, right, and left). If the worm finds a square with a number it has never encountered, it enters that square. The worm then starts again, scanning in four directions for a potential move from its new location. The process continues until the worm can no longer find a "safe" square, at which point its meager life comes to an end. When I discussed this with colleagues, we imagined that a worm spends a year on each square, at which point it grows a foot in length. Therefore, a worm that has passed through ten squares before going into anaphylactic shock and dying is ten years old and 10 feet long.

   Table 13.2 (next page) is sample output from Program 1 showing the longest worm found for each starting square in Worm World 1. You can see that a variety of worms are produced. For World 1, worms range in size from 5 feet (starting in square 29) to a humongous 20-foot-long monstrosity that started in square 2. Large, ancient worms almost as long as 20 feet occur fairly often, according to Table 13.2. (Note that the longest known species of earthworm is *Microchaetus rappi* of South Africa. In 1937 a giant measuring 22 feet in length was found.)

   So you can get an idea of the range of worm sizes, Figure 13.1 shows a plot of all the worm lengths as a function of starting square for all three Worm Worlds. The youngest dead worm found so far by the computer program is 3 years old. The oldest is 22 years old. Notice that the plots are not random—the distribution of worm ages (lengths) clusters in interesting ways.

The Internet Worm World Tournament

In the previous section we saw that the computer programs were capable of finding some ancient worms, but are there older worms lurking in Worm World? After all, the computer programs did not search for every possible path a worm could take. For example, once a worm found an available square, it would commit itself to moving

TABLE 13.2.   Some Worm Paths in Worm World 1

```
(1 15)*   0   4  22  24  17  12   5  15  11   3  23   6  19  18  16
(2 20)    0   2  21  17  12   4  11  19   8   6  15  24  18  16   5   7  20   1   3  13
(3 19)    2  21  17  12   4  11  19   8   6  15  24  18  16   5   7  20   1   3  13
(4 13)   21   2  23  15  18  19  11   4  12  17  24  22   3
(5 12)    2  23  15  18  19  11   4  12  17  24  22   3
(6 11)   23  15  18  19  11   4  12  17  24  22   3
(7 12)   15  18  19  11   4  12  17  24  22   3  23   6
(8 12)   18  15  23   2  21  17  12   4  11  19   8   6
(9 14)    4  22  24  17  12   5  15  11   3  23   6  19  18  16
(10 17)  22  24  17  12   4  11  19   8   6  15  14  18   7  20   1   3  13
(11 16)  24  17  12   4  11  19   8   6  15  14  18   7  20   1   3  13
(12 17)  17  12   4  11  19   8   6  15  24  18  16   5   7  20   1   3  13
(13 16)  12   4  11  19   8   6  15  24  18  16   5   7  20   1   3  13
(14 16)   4  11  19   8   6  12   5  15  17  24  22   3  23  18   9   7
(15 17)  11  19   8   6  12   5  15  17  24  22   4   3  18  23   7   9   2
(16 10)  19  11   4  12  17  24  22   3  23   6
(17 7)    3  18  23   6  12  15   5
(18 9)    3  11  15   5  12   6   8  19  18
(19 8)   11  15   5  12   6   8  19  18
(20 15)  15   5  12   6   8  11  18  14  20   7   3   1   0  23  21
(21 9)    5  12   6   8  11  15  24  18  16
(22 14)  12   6   8  11  15  24  18  16   5   7  20   1   3  13
(23 7)    6   8  11  15  24  18  12
(24 15)   8   6  12   5  15  11   3  23  18   9   7  19  10  17   2
(25 6)   18  23   6  12  15   5
(26 9)   23   6  12  18  24  15  11   8  19
(27 8)    6  12  18  24  15  11   8  19
(28 7)   12  18  24  15  11   8   6
(29 8)   18  24  15  11   8   6  12   5
(30 19)  24  15  11  18  14  20   7   3   1   0  23  21   8   6   5  10  17   2   9
(31 15)  15  11  18  14  20   7   3   1   0  23  21   8   6   5  24
(32 13)  11  15  24  18  12   6  23   7  19  10  17   2   5
```

(continued)

in that direction without examining *every* possible path it could take. (This is a little like real life, is it not?)

To see whether humans could beat the computer results, I held a grand Internet tournament in which I asked various interested colleagues on the computer networks to find the oldest worm in each of the three Worm Worlds. I also asked whether it is

TABLE 13.2.   (Continued)

(33 10)	9	7	19	18	16	5	6	10	17	2						
(34 10)	7	19	18	16	5	6	10	17	2	9						
(35 9)	19	18	16	5	7	20	1	3	13							
(36 8)	18	16	5	7	20	1	3	13								
(37 14)	16	18	14	20	7	3	1	0	23	21	8	6	5	24		
(38 13)	18	14	20	7	3	1	0	23	21	8	6	5	24			
(39 7)	14	18	7	20	1	3	13									
(40 13)	18	14	20	7	3	1	0	23	21	8	6	5	24			
(41 15)	2	17	10	6	5	7	20	1	3	13	18	23	21	8	24	
(42 14)	17	10	6	5	7	20	1	3	13	18	23	21	8	24		
(43 13)	10	6	5	7	20	1	3	13	18	23	21	8	24			
(44 12)	6	5	7	20	1	3	13	18	23	21	8	24				
(45 11)	5	7	20	1	3	13	18	23	21	8	6					
(46 12)	7	20	1	3	13	18	23	21	8	6	5	24				
(47 12)	20	7	3	1	0	18	24	5	6	8	21	23				
(48 12)	7	20	1	3	13	18	23	21	8	6	5	24				
(49 5)	5	24	21	6	8											
(50 16)	24	5	6	8	21	23	18	13	3	1	0	7	20	14	15	11
(51 7)	5	24	18	0	1	3	13									
(52 6)	24	18	0	1	3	13										
(53 5)	18	0	1	3	13											
(54 11)	0	1	3	13	18	23	21	8	6	5	24					
(55 10)	1	3	13	18	23	21	8	6	5	24						
(56 10)	3	1	0	18	24	5	6	8	21	23						
(57 4)	24	21	6	8												
(58 9)	21	6	8	24	18	0	1	3	13							
(59 14)	6	8	21	23	18	13	3	1	0	7	20	14	15	11		
(60 13)	8	21	23	18	13	3	1	0	7	20	14	15	11			
(61 16)	21	23	18	13	3	1	0	7	20	14	15	11	8	6	12	5
(62 15)	23	18	13	3	1	0	7	20	14	15	11	8	6	12	5	
(63 11)	18	13	3	1	0	23	21	8	6	5	24					
(64 8)	13	18	23	21	8	6	5	24								

*The numbers in parentheses are the starting square number and final path length for each square in Worm World 1. These were computed using the C program.

possible to design a Worm World in which each starting square would yield the same maximum length. Many respondents used pencil and paper to investigate the worlds.

James Wicht, a mathematics major and freshman at Case Western Reserve University in Cleveland, was one contestant who did a good deal of searching by hand. On the next page are the oldest, largest worms he could find. Squares on the worm path are indicated by an asterisk to the right of the number.

0	0	2	21*	2*	23	15	18
4	22*	24*	17*	12*	4*	11*	19
3	3*	11	15	5	12	6*	8
18	23*	6	12	18	24	15*	11
9	7*	19*	18	16	18	14*	18
2	17	10*	6	5	7	20*	4
5	24	5*	24	18	0	1*	3
24	21	6	8	21	23	18*	13*

12	7*	11*	4	22	22	17	6
15	24*	1*	12*	9*	8*	16	20
11	23*	15	9	17	6*	24	4
23	21*	3*	4*	8	17*	22*	17
6	6	22*	16*	2	5*	8	1
11	10	17*	2	8	18*	14*	12
7	11	24*	3*	13*	10*	15*	7
20	20	3	1	9	7	10	6

2	13	20	9	19	8	16	12
22	15	116	14	14	11	12	5
13	12	3	6	4	12*	8*	15
4	21	8	18	17	3*	5*	1*
9	14	15	19	15	4*	4	2*
16	6	16	4	11	16*	3	17*
3	15	15	3	15*	14*	8	21*

The method James used to determine these paths is straightforward. First he located all the squares with unique numbers and marked them. He also marked numbers that appeared many times, as these limited the length of possible worm paths. Next, he searched for a path that went through a group of unique numbers or was the only way through an area with many repeated numbers. He then tried to extend the path in both directions. In some cases, he would skip the opportunity to visit a unique number if it was near a corner or near frequently occurring numbers. Wicht required about 8 minutes of hand searching for each Worm World. He notes that extremely short worms may occur if recurring numbers surround a small group (as in the low 8 in Worm World 3).

Scott Purdy, a Pomona College freshman, writes, "I believe that the longest possible path for Worm World 1 is 24 spaces, omitting only the solitary 16." Starting at the lower right, the path he found is 13, 18, 1, 20, 14, 15, 6, 8, 9, 11, 4 12, 17, 21, 2, 0, 22, 3, 23, 7, 19, 10, 5, 24. He says, "Any method I can see for connecting the solitary 13 to the solitary 16 prevents me from getting to the solitary 22. As you might have guessed, I began with the singletons and tried to connect as many as possible. All those 18's near the center of the board and around the 16 told me I was in trouble there." For Worm World 1, Scott believes that the square containing the 16 would produce the shortest paths because of the number of 18's around it.

The most rigorous treatment of the problem came from Stephen Tavener of the United Kingdom. Stephen is a 26-year-old computer programmer/team leader for Logica Company with a degree in pure mathematics from Queen Mary College, London University. He spent an hour creating a recursive C program (see Appendix 2) that required about a minute per square to determine the longest and shortest paths. (A recursive program is one that calls itself, and a recursive function is on that is defined in terms of itself.)

When Stephen used his recursive program to perform an exhaustive search for longest paths in Worm World 1, he found a range of path lengths as follows (R=right, L=left, U=up, D=down):

Minimum: 18, maximum: 24

Longest sequence from (0, 1) is RDRRDLDLLDRRDRDRRURUULL (length: 24)

Longest sequence from (4, 4) is RRDDLDLLULULLURRURULLUL (length: 24)

Longest sequence from (0, 7) is RRUULLURRRUULLULU (length: 18)

Notice that there are two length-24 paths. The paths for Worm World 2 are:

Minimum: 18, maximum: 23

Longest sequence from (0, 0) is RRDLDDDRRURRDDRDLLDLLL (length: 23)

Longest sequence from (6, 0) is RDLLLULDLLULDDRDR (length: 18)

Longest sequence from (1, 7) is RRRURRULUULLDLLUUURULL (length: 23)

The paths for Worm World 3 are:

Minimum: 17, maximum: 21

Longest sequence from (2, 0) is RDDRRDRRDDDDLLLLURUR (length: 21)

Longest sequence from (3, 0) is LDRDRRDRRDDDDLLLLURU (length: 21)

Longest sequence from (6, 0) is LLLLLLDRDRRDRUUR (length: 17)

Longest sequence from (2, 1) is URDDRRDRRDDDDLLLLURU (length: 21)

Longest sequence from (4, 5) is DLDRRRRUUUULLULLULUR (length: 21)

Longest sequence from (5, 5) is LDLDRRRRUUUULLULLUUL (length: 21)

Stephen Tavener also designed two "extreme" worlds. The first one has a maximum path length of 25 from all starting points!

0	1	2	3	4	5	6	7
15	14	13	12	11	10	9	8
16	17	18	19	20	21	22	23
6	5	4	3	2	1	0	24
7	8	9	10	11	12	13	14
22	21	20	19	18	17	16	15
23	24	0	1	2	3	4	5
13	12	11	10	9	8	7	6

The following board has a maximum length of 1 from all starting points:

0	0	0	0	0	0	0	0
0	0	0	0	0	0	0	0
0	0	0	0	0	0	0	0
0	0	0	0	0	0	0	0
0	0	0	0	0	0	0	0
0	0	0	0	0	0	0	0
0	0	0	0	0	0	0	0
0	0	0	0	0	0	0	0

He also constructed a Worm World where the maximum length was less than or equal to 2 from all points and the world contains as many distinct numbers as possible:

1	0	0	2	0	0	0	3
0	0	0	0	0	4	0	0
0	5	0	0	0	0	0	0
0	0	0	6	0	0	0	7
8	0	0	0	0	9	0	0
0	0	10	0	0	0	0	0
0	0	0	0	0	0	0	0
11	0	0	12	0	0	0	13

Abkhasian Areas

Abkhasia, a remote mountain region near southern Russia, seems to produce humans of extreme longevity. In fact, Abkhasians have a distinct word for great-great-great-grandparents that is applied only to the living. Are there Abkhasian areas in Worm World? Do certain starting squares in the chessboard have a higher probability of yielding older worms? For example, do interior squares generally yield longer possible worm paths than border squares because they have more neighbors? All these questions are as yet unanswered, though it would seem possible to empirically find such regions by having the computer simulate thousands of Worm Worlds and then coloring those squares that produce older worms red. Regions with young dead worms might be blue. Do any correlated regions of color arise after many simulations?

James Wicht believes that border and near-border squares are fertile breeding grounds for long worms. He suggests that worms going through the center of the world develop restrictions on their path because there are so many more ways to "bump into" repeated allergens. "Certainly," he says, "at least the ends of the paths should be near corners."

Similarly, Stephen Tavener believes that at least one end of any path should be near or on the edge of the board. He says, "If you go snaking off across the middle of the board, you're probably reducing the reachable squares. Optimum paths are likely to be more-or-less spirals from the middle which go out to the edge, or squiggles working out from the vicinity of a corner."

The Czech Logical Labyrinth

About a month after I posed the Anaphylactic Worm Game to colleagues, I came across a very similar looking game from the Czech Republic. It was called a logical labyrinth. For an entire week, the city of Brno hosted 52 competitors from 13 countries in the 2nd World Puzzle Championship. The logical labyrinth (below) was just one of many puzzles the participants had to solve.

1	2	7	16	5	19	7	8
10	17	23	9	21	13	6	23
15	4	22	17	20	4	18	22
19	17	11	24	3	17	8	19
12	9	7	14	23	18	12	6
6	5	23	14	10	5	2	23
8	20	10	11	19	18	1	17
2	24	15	22	13	4	4	25

The goal is to find a path from the 1 (upper left) to the 25 (lower right) such that the path contains exactly 25 squares. As in my Worm Worlds, the moves are up, down, right, and left, and the path may not repeat any numbers. In the Czech tournament, contestants had only 15 minutes to solve the problem, and computers were not allowed. Can you find the path? Incidentally, the countries' scores in the tournament were, from highest to lowest: Czech Republic, United States, Canada, Poland, Japan, Turkey, Hungary, Slovakia, Argentina, Croatia, Netherlands, Slovenia, and Germany.

Hyperworms in Abkhasia

Worm World has just begun to be explored, and there is fertile ground for future investigations. I am interested in hearing from readers who have explored some of the areas in the following list.

1. Extend the simulation to three-dimensional (3-D) worlds where worms tunnel through a cube of "Swiss cheese." Use computer graphics to display the longest paths as the computer finds them.
2. Explore large Worm Worlds. To this date, only 8-by-8 worlds have been explored. Why not try something incredibly challenging, such as a 20-by-20 world? Stephen Tavener notes that larger worlds will require so much computer resources that his recursive program will probably crash when trying to solve boards larger than 15-by-15 with their correspondingly greater range of numbers. He says that "larger worlds would have forced me to do some real thinking." He also suggests a game in which the sequence of numbers along a worm path must always be increasing.

3. Work with multiple worms all competing for space. Challenge your friends to compete with you on the same world.

4. If you were given a single dose of antihistamine so the worm could, at one time during its life, go through a cell with a repeated allergen, how would this affect your simulations?

5. Search for Abkhasian areas in 3-D simulations. Use hyperworms in four-dimensional (4-D) simulations, and search for hyperdimensional Abkhasian areas there as well.

6. Given Worm Worlds constructed randomly as in this chapter, what is the *average* largest worm path you would expect to find? Chris Lusena from Ontario, Canada, ran 100 simulations with even probability for each allergen in the grid and found that the average longest path is around 21 squares. The average shortest path is around 17 squares. How do these averages scale with the size of the board?

7. How would you expect the game to change if it were to be played on worlds the size of our Earth or solar system?

In closing, I give the answer to the Czech logical labyrinth: 1, 10, 15, 4, 22, 11, 7, 14, 24, 3, 20, 21, 9, 16, 5, 19, 13, 6, 18, 8, 12, 2, 23, 17, 25.

Worm World constructed with rings. Insert numbers on the various rings and see how far your worms can travel using the rules in this chapter.

CHAPTER 14

Consider the true picture. Think of myriads of tiny bubbles, very sparsely scattered, rising through a vast black sea. We rule some of the bubbles. Of the waters we know nothing . . .

— Larry Niven and Jerry Pournelle, *The Mote in God's Eye*

Hiding between all the ordinary numbers was an infinity of transcendental numbers whose presence you would never have guessed until you looked deeply into mathematics.

— Carl Sagan, *Contact*

Fractal Milkshakes and Infinite Archery

Imagine a frothy milkshake with an infinite number of bubbles. As you stare into the cosmological milkshake, you notice bubbles of all sizes, touching one another but not interpenetrating. In this hypothetical foam, the bubbles become smaller and smaller, always filling in the cracks and spaces between larger bubbles. If you were to magnify the foam, tiny bubbles would always be interspersed with larger ones, but the overall structures would look the same at different magnifications. In other words, the froth would be called a fractal because it displays self-similar structures at different scale sizes.

Such thoughts of an infinite froth need not be confined to mere armchair speculation. In fact, we can construct and explore such a froth that inhabits the realm of abstract geometry. Moreover, the froth we will make is not simply an artistic design created from some arbitrary mathematical recipe, but rather it comes about naturally from the study of the little-known Ford circles, named after L. R. Ford, who published on this topic in 1938. Ford circles provide an infinite treasure chest to explore, and the circles are among the most mind-numbing mathematical constructs to contemplate. In fact, it turns out that they describe the very fabric of our "rational" number system in an elegant way.[1]

Recall that rational numbers are numbers that can be expressed as fractions. For example, 1/2, 1/3, and 2/3 are all rational numbers. As you might expect, an infinite number of such numbers inhabits our mathematical universe.

What follows is a mathematical recipe for creating a Ford froth that characterizes the location of rational numbers in our number system. You can use a compass and some graph paper to get started. No complicated mathematics is required for your journey. Let us begin by choosing any two integers, h and k. Draw a circle with radius $1/2k^2$ and centered at $(h/k, 1/2k^2)$. For example, if you select $h = 1$ and $k = 2$, you draw a circle centered at $(0.5, 0.125)$ and with radius 0.125. Note that the larger the denominator of the fraction h/k, the smaller the radius of its Ford circle. Choose another two values for h and k, and draw another circle. Continue placing circles as many times as you like.

As your picture becomes more dense, you will notice something quite peculiar. None of your circles will intersect, although some will be tangent to one another (just kissing one another). Figure 14.1 is a computer graphics rendition of the froth where I plotted 225 circles corresponding to 225 different fractions. In particular I let h and k assume values 1 through 15. (The program code in Appendix 1 gives you additional details.) Even if we place infinitely many Ford circles, none will overlap,

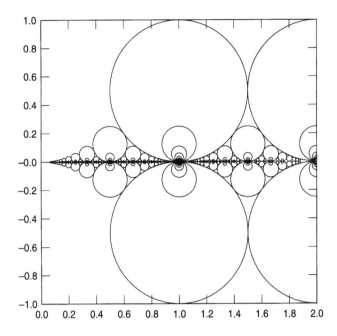

FIGURE 14.1.   Fractal Ford froth.

and each will be tangential to the *x*-axis. We can get a visual confirmation of this by magnifying the froth. Figure 14.2 is a magnification of the left-most tip of the froth, further revealing its fractal structure. Would this image look very different if I had increased the number of different fractions represented on the graph?

Ford circles remind me of some of the surreal drawings by Doré, specifically those drawings of infinite collections of angelic circular rings illustrating Dante's *Divine Comedy* (Figure 14.3). I first learned about Ford circles from the late Hans Rademacher (1892–1969), who lectured on them at Stanford University in 1947 and whose lecture material is reprinted in his 1983 book *Higher Mathematics from an Elementary Point of View*. Unfortunately there were no computer graphics back in 1947.

When I showed colleagues my computer graphics of Ford-circle froth, they were at first confused by the occasional instances of circles within circles. (In fact, the only other clear diagrams I had seen of Ford circles were in Rademacher's mathematics text, and the figures were hand-drawn diagrams containing only about 13 circles.) These anomalous interior circles, which I call Chrysler circles, arise from the fact that my computer programs produce various h/k values that are equivalent, such as 1/2, 2/4, 4/8, ... This, in turn, produces circles with different radii but having the same x coordinate. In addition, I also plot circles at $(h/k, -1/2k^2)$ to increase the symmetry of the representation and heighten its aesthetic appeal.

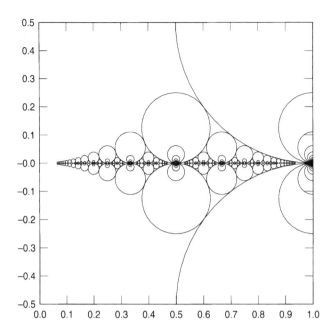

FIGURE 14.2.   Fractal Ford froth, magnification of Figure 14.1.

How many neighbor circles touch an individual circle? For the mathematically inclined reader, note that two fractions are called adjacent if their Ford circles are tangent. Any fraction has, in this sense, an infinitude of adjacents. Any circle has an infinitude of circles that kiss it.

Archery in the Great Beyond

God is an intelligible sphere whose center is everywhere and whose circumference is nowhere. — Hermes Trismegistus

Do you recall the scene in the Robin Hood story where Robin Hood's aim is so precise he is able to shoot his arrow right through the center of another arrow that is lodged in a tree? For a moment, let us pretend we are all are as good as Robin Hood—or perhaps a bit better.

Consider a godlike archer who launches an arrow at the Ford froth. Depending on where the archer aims, the outcome has very exciting and different outcomes. To understand this, place your virtual archer high up; that is, select a position above the Ford froth with an appropriately large y value. To simulate the shooting of the arrow, next draw a vertical line from the location of your archer (for example, at $x = a$). Trace the arrow's straight-line trajectory as gravity pulls it down to the x-axis. It

FIGURE 14.3.   Infinite circlular rings comprised of angels illustrating Dante's *Divine Comedy*.

turns out that if the archer's position is at a rational point (a is rational), the line must pierce some Ford circle and hit the x-axis exactly at the circle's point of tangency. However, when the archer's position is at an *irrational* number (a nonrepeating, endless decimal value such as $\pi = 3.1415 \ldots$), the arrow cannot pass directly to the x-axis from a Ford circle. In other words, the arrow must *leave* every circle it enters. However, as I mentioned previously, every circle it leaves is completely surrounded

FIGURE 14.4.   3-D representation of the froth. The color of the spheres is related to their radii. If desired, transparency can be used to visualize some of the internal Chrysler spheres with equivalent h/k ratios.

by a chain of adjacents. Therefore the archer's arrow travelling along $x = a$ must enter another circle! This is true for all the Ford circles the arrow pierces. Therefore, when the godlike archer is located at an irrational point, the archer's arrow must pass through an infinity of circles!

This all relates to the fact that even though there are an infinite number of rational and irrational numbers, the infinite number of irrationals is in some sense greater than the infinite number of rationals. To denote this difference, mathematicians refer to the infinity of rationals as $\aleph_0$ and the infinity of irrationals as C, which stands for continuum. There is a simple relationship between C and $\aleph_0$. It is $C = 2^{\aleph_0}$. The continuum hypothesis states that $C = \aleph_1$; however, the question of whether or not C truly equals $\aleph_1$ is considered undecidable. In other words, great mathematicians such as Kurt Gödel proved that the hypothesis was a consistent assumption in one branch of mathematics. However, mathematician Paul Cohen proved that it was also consistent to assume the continuum hypothesis is false!

Interestingly, the number of rational numbers is the same as the number of integers. The number of irrationals is the same as the number of real numbers. (Real mathematicians usually use the term *cardinality* when talking about the "number" of infinite numbers. For example, true mathematicians would say that the cardinality of the irrationals is known as the continuum.)

In Figures 14.4, 14.5, and 14.6, I have represented the Ford circles as spheres and displayed them from different viewpoints and magnifications. By using transparency, we can gaze into some of the spheres and see the internal spheres produced by identical h/k ratios for different values of h and k. Color is used to indicate

FIGURE 14.5.   Magnification of Figure 14.4.

changes in the values of the circles' radii. You can think of the two-dimensional (2-D) figures as cross-sections of the 3-D figures. Is it possible to use three integers, h, k and l, to expand the 3-D plot further into the z direction?

Figure 14.7 shows another way to characterize the distribution of rational numbers, which I have not seen presented before. Each circle is located at $(h/k,k/h)$ with

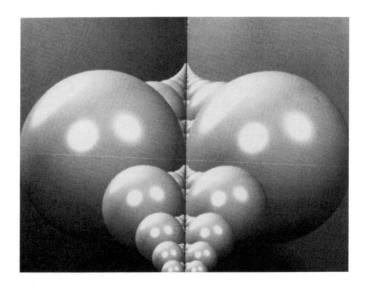

FIGURE 14.6.   A different view of Figure 14.5.

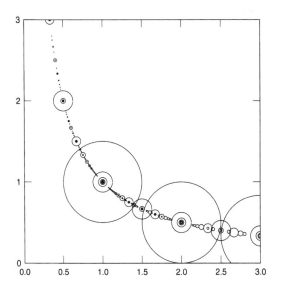

FIGURE 14.7.   An $(h/k, k/h)$ plot.

radius $1/2k^2$. What happens if we take a journey along the y-axis? What structures will we find higher up along the graph? Figure 14.8 is a magnification of Figure 14.7.

Ford froth is endless. With appropriate computer programs, students, artists, and mathematicians can "swim" through the froth like a fish through the surf. What

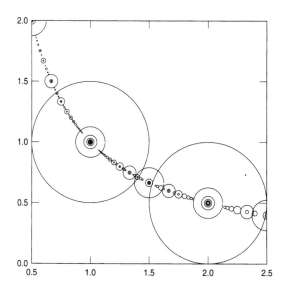

FIGURE 14.8.   Magnification of Figure 14.7.

strange oceanic worlds will you find as you explore different regions of the fractal Ford froth?

Further Exploration

Gary Adamson from California suggests that we do not plot Ford circles for "every" possible rational number, but rather that we can confine ourselves to a specific subset of fractions. As one example, consider the sequence of fractions that eventually converge to the golden ratio,

$$\varphi = (\sqrt{5} + 1)/2$$

These fractions are:

$$\frac{1}{1}, \frac{2}{2}, \frac{3}{2}, \frac{5}{3}, \frac{8}{5}, \cdots \rightarrow \frac{\sqrt{5} + 1}{2}$$

(Note that denominators and numerators are formed from the Fibonacci sequence, in which each term is the sum of the previous two terms.)

What would golden ratio Ford circles look like? How many tangent neighbors does each circle have? How many circles (fraction terms of the sequence) can you represent before the smallest is too small to be seen without further magnifying your picture?

A Musical Tribute

When Kevin D. Quitt from NASA's Jet Propulsion Laboratory heard of my interest in the different orders of infinity, he posted the following lyrics to a mathematical newsgroup. I couldn't resist reprinting it here, with his permission:

$\aleph_0$ bottles of beer on the wall,

$\aleph_0$ bottles of beer.

Take one down,
Pass it around,

$\aleph_0$ bottles of beer.

Postscript

Some technical readers may like to see a proof that the Ford circles representing any two different fractions cannot overlap. (In the extreme case, they may be tangent.) The mathematical argument follows that of Rademacher.

Let us consider two Ford circles represented by different fractions h/k and H/K. If d is the distance between their two centers, we have

$$d^2 = (H/K - h/k)^2 + (1/2K^2 - 1/2k^2)^2$$

(It helps to see this if you draw a diagram showing the two circles.) The sum of the two circles' radii is $s = r_1 + r_2 = 1/2k^2 + 1/2K^2$. Next, it is helpful to compare d^2 with the square of the sum of the radii, s^2. With some algebraic manipulation we find

$$d^2 - s^2 = ((Hk - hK)^2 - 1)/k^2K^2$$

We also find that $Hk - hK \neq 0$ because h/k and H/K are different fractions.

It turns out that $(Hk - Kh)^2 \geq 1$, so that $(Hk - Kh)^2 - 1 \geq 0$, or $d^2 - s^2 \geq 0$. This means that the distance between the centers of circles h/k and H/K is greater than or, at minimum, equal to the sum of the radii, in which case the circles are tangent. The two circles will be tangent provided $|Hk - Kh| = 1$. If you wish to compute all the adjacent circles H/K to a particular h/k circle, use the following formula:

$$H_n/K_n = (H + nh)/(K + nk)$$

where $n = 0, \pm1, \pm2, \ldots$

C H A P T E R **15**

"*These are the main simulation computers,*" *Copernick said.* "*Each of these can simulate the entire life-cycle of an organism. With a fifty-gigahertz clock, I can take a human being from a fertilized cell to an octogenarian in eleven hours. They are our most important single tool we use in bioengineering.*"

— Leo A. Frankowski, *Copernick's Rebellion*

Creating Life Using the Cancer Game

Creation

Since its discovery around 1869, deoxyribonucleic acid (DNA) has emerged as the most scintillating star in the universe of biological molecules. DNA contains the basic genetic information of all living cells. Today we know that its information is coded by a sequence of bases symbolized by A, C, G, and T, standing for adenine, cytosine, guanine, and thymine. The bases code information concerning protein synthesis as well as a variety of regulatory signals.

DNA is a large molecule supported by a twisting sugar-phosphate backbone, and it contains millions of such bases. The DNA strings in a single cell would measure up to 6 feet in length if stretched out, but an elaborate packing scheme coils the strings to fit within a cell only 1/2,500 of an inch across. Special enzymes copy the genetic information. Additional enzymes check to make sure the copying process is correct. You might think of these enzymes as zealous police inspectors, and as a result of their meticulous patrol, the genetic data processing makes only about one error in a billion copy steps.

Interestingly, ribonucleic acid (RNA), a molecule very similar to DNA, is thought by some to be the most primitive "life" that first evolved: The strands containing the sequences spontaneously fold into complex two-dimensional structures, and reproduce given the right conditions. (Like DNA, RNA uses four bases, except that the base U (uracil) replaces T.)

You may be interested in studying a simulation that is a very simple artificial genetics model. I call this game the Cancer Game because the simple genetic rules can produce both stable behavior and uncontrolled growth of the DNA segments. You can also think of the simulation as a Creation Game as you watch ever-larger and complicated sequences arise from two tiny progenitor sequences. The simulator program uses the letters G, C, A, and T to represent the four chemical bases, or nucleotides, that make up DNA. The period of chemical evolution on Earth, whereby organic compounds gradually accumulated in the primitive seas, probably began about 4,000 million years ago. Our experiments will be much quicker.

The Game

Consider the following Cancer Game, which makes use of a computer enzyme to create complicated genetic sequences. Amazing complexity arises in just a few generations. Try the BASIC or C simulator programs in Appendix 1, and make modifications to grow your own genetic sequences.

To start the game, a "polymerase enzyme" is used to create a new sequence from old ones. The enzyme has two regions: a copy part (called COP) and a mutant part (called MUT). COP's job is to make a copy of all the bases in the previous generation. MUT's job is to make a copy of all the bases in the second-to-last generation, but mutate them so that G and A are switched and C and T are switched whenever they occur. The whole enzyme then links all the nucleotides together.

Here is an example. Start with a G for the first generation and a C for the second. Now add the polymerase enzyme. COP copies the C and places it as the last base in generation 3. MUT takes the G, converts it to an A, and then attaches it to the C, producing AC. So far we have:

Generation	Nucleotide Sequence
1	G
2	C
3	AC

Next, the enzyme continues to make a longer strand. It "examines" generations 2 and 3, producing:

4	TAC

And so on:

5	GTTAC

Up to this point I have not mentioned the process of sequence cleavage because it did not apply. We now must consider the splitting enzyme, CLEAVE. Whenever two T's occur in a row, CLEAVE splits the sequence, and the generating process continues in the left branch as before. (This splitting is a metaphor for enzymatic cleavage at a particular recognition pattern.) The dash indicates a splitting has occurred:

5	GT-TAC	(split at TT)

(The TAC fragment is split off and floats around in the soup. It does not contribute to any further growth process.)

6	CGTGT	(combining MUT(4) and the GT from 5)
7	ACCGTGT	(combining MUT(5) and the CGTGT from 6)

The process continues for as long as you like.

I asked colleagues a number of interesting questions concerning the game, and I am sure you can think of many more.

1. What is the longest genetic strand produced by this process in 1,000 generations?
2. Does the system grow out of control and rapidly produce superlarge chunks (cancer), or does it limit itself to small chunks?
3. Is it possible for the strands to grow larger and larger for a few generations and then to disintegrate into strands of only one or two bases?
4. What is the smallest sequence floating in the soup?
5. Will an isolated triplet of three identical nucleotides ever evolve in the fragment soup, such as GGG or CCC?
6. Will a triplet of three identical nucleotides ever evolve in any sequence?

One of the easiest ways to study the system is to generate the sequences using a computer program. To start you on the task, and to provide the seed for more involved scenarios, I have provided both C and BASIC code in Appendix 1. Table 15.1 shows the lengths and sequences for the first 19 generations. The behavior of this sequence is quite interesting. The lengths start slowly growing in size until a sequence of length 19 is achieved in generation 9. After this, the sequences seem to undergo a high-frequency warble. In other words, there seems to be an oscillation of strand lengths. The beginning of the warble is clearly seen in Figure 15.1. Does the warble continue? The sequences do get pretty bizarre looking. For example, the 52nd generation is:

TACACACACACACACACACACACACACACACCGTGTGTGTGTGTGTGT-
GTGTGTGTGT

It is possible to make a number of interesting observations on the sequences. For example, the largest string found after searching the first 1,000 generations contains 1,507 bases. Also, if we remove the cleavage enzyme so the sequences never split, the length of successive sequences follows the Fibonacci sequence. The Fibonacci sequence of numbers (1, 1, 2, 3, 5, 8, ...) plays important roles in mathematics and nature. These numbers are such that, after the first two, every number in the sequence equals the sum of the two previous numbers, or $F_n = F_{n-1} + F_{n-2}$. If CLEAVE is used, another observation is that every sequence after generation 5 terminates with a T. Also, after thousands of generations, the smallest fragment found floating in the soup is TAC.

Various researchers have studied the Cancer Game, including David Epstein, a member of the Department of Information and Computer Science at the University of California, Irvine. He notes that the starting character repeats every four generations. He also believes that the "after-split" sequence lengths, starting in generation 6, can be described by the recurrences

$$L(4i) = L(4i - 1) + L(4i - 2) - L(4i - 2) = L(4i - 1)$$
$$L(4i + 1) = L(4i) + L(4i - 1)$$
$$L(4i + 2) = L(4i + 1) + L(4i)$$
$$L(4i + 3) = L(4i + 2) + L(4i + 1)$$

TABLE 15.1.   The First 19 Sequences

Length	Sequence
1	G
1	C
2	AC
3	TAC
5	GTTAC
5	CGTGT
7	ACCGTGT
12	TACACACCGTGT
19	GTTACACTACACACCGTGT
14	CGTGTGTTACACGT
9	ACCGTGTGT
16	TACACACACCGTGTGT
25	GTTACACACTACACACACCGTGTGT
18	CGTGTGTGTTACACACGT
11	ACCGTGTGTGT
20	TACACACACACCGTGTGTGT
31	GTTACACACACTACACACACACCGTGTGTGT
22	CGTGTGTGTGTTACACACACGT
13	ACCGTGTGTGTGT

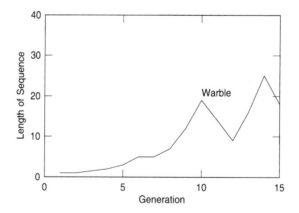

FIGURE 15.1.   The beginning of the high-frequency warble in genetic strand lengths.

with initial conditions $L(5) = 2$ and $L(6) = 5$. This recurrence defines a modified Fibonacci sequence where every fourth position repeats the previous one. Finally, David discovered that there will never be triplet of identical bases, such as CCC or AAA.

Dean Hickerson from the University of California, Davis, notes that generation $4n$ can be symbolically represented as $T(AC)^{n+1} C(GT)^n$, where the superscripts indicate that the small sequences in parentheses repeat a specified number of times. He also finds that generation $4n + 2$ can be represented as $C(GT)^{n+1}$ and that generation $4n + 3$ can be represented as $ACC(GT)^{n+1}$. (These observations do not apply to the first four generations.)

Genetics, Programs, and Evolution

Our genetic code is similar to a large computer program. About 6 billion bits are used to describe a human, in contrast to tens of millions of bits in very complicated computer programs. Using the rules given in this chapter, how much time would it take a computer program to generate a genetic sequence as large as a human's? What is the relative frequency of all the different species floating in the primordial soup after many generations? Are there dominant sequences? Would other seed sequences lead to more interesting behavior?

You may wish to modify the Cancer Game so it becomes more of an Evolution Game in which the different segments of genetic material may interact or combine in interesting manners. In addition, a graphics interface whereby the genetic pieces actually code for computer graphics representations of distinct primitive organisms may allow researchers to look at the primordial soup to determine how the evolution process is progressing. I hope you share my pleasure in studying a number of these systems starting with different sequences and introducing different mutation rules.

A 40-Day Flood

While pondering the mysteries of living systems and their growth and reproduction, I asked a number of colleagues the following questions:

1. Given a large container enclosing two individuals of every animal species in the world, what would be the approximate total weight of all the organisms? Would this weight be more than the weight of the Empire State Building in New York City? How would your answer differ if you included every plant, bacterial, and fungal organism?

2. Assume that all other organisms on Earth were dead except for those pairs in the container in question 1, and assume that the animals were released 1,000 years ago. What would you expect to be surviving today? (Assume that, where applicable, a male and female were used for each species.)

3. Assume that today it started to rain for 40 days, and the rain covered all the land on the Earth. Further assume that the flood waters receded to preflood days within several months. What would be the geopolitical changes as a result of the temporary flood? What would be the ecological changes?

I received a number of fascinating replies from scientists. Biologist Ralf Stephan estimated the weight of animals in the container to be 1,000 tons. He used a value of 10 million for the number of species and assumed an average mass of 100 grams. (Insects decrease his figure for average mass because of the huge number of insect species.) He also says there would be no significant increase in mass as a result of including bacteria or fungi, but there would be some increase if plants were used. (How would this change if extinct species were included?) Other scientists investigated the weight problem by working in units of blue whales, the "champions in the weight department." Lou Bjostad from Fort Collins, Colorado, estimated the weight of all the animal pairs in the enclosure to be about ten whales' worth.

Surprisingly, most biologists said that no animals would be alive after 1,000 years because approximately 50 individuals of a single species are needed to sustain genetic health. There is also the additional problem of making sure that both male and female offspring are surviving. Today, species are considered endangered well before their numbers drop below 50.

Ralf Stephan also asked for clarification of my phrase "rain covered all the land." If the phrase means that all the land including the Himalayas is covered evenly, this would require a depth of 9,000 meters in most regions. Further, he asked how fast will the water rise, and whether there is time to put people and food into boats. He indicated that most people do not have access to a boat and that most airplanes could not land, although some might float. The most important variable for immediate survivability depends on the distribution of floating objects, because if people must swim 500 to 1,000 meters in cold water to reach a floating object, most will drown.

One scientist indicated that such an inundation would probably also kill most of the plant life on Earth. Even if the waters were to recede, the resultant salt deposits would prevent plants from growing for many years. The scientist suggested that, during these years, many of the seeds and plants that could live through the flood itself would die out.

Other scientists wondered about the ecological effect of the numerous dead carcasses caused by the initial flood. Some biologists indicated that the initial health of each pair of organisms should be defined in the original question. Another variable to consider is whether the animals are all released in one place as opposed to being spread evenly over the planet.

One colleague noted that all the pairs of animals require food. "If one animal of species A eats one of species B, then B becomes automatically extinct (unless the male was eaten, and the pair had already initiated reproduction)."

Andy Symes wrote that he did not know what geopolitical changes would take place as a result of the rain, but that it was "a safe bet that the Netherlands *wouldn't* get flooded." He said, "We've been keeping the sea out for hundreds of years, so we've more experience at this problem than anyone else."

CHAPTER 16

One thing I know and that is that I know nothing.

— Socrates

No Zeros Allowed

What is special about 1 decillion? For one thing, 1 decillion is a very large number: 10 raised to the power of 33, or 1 followed by 33 zeros:

1,000,000,000,000,000,000,000,000,000,000,000

Aside from its obvious enormity, there is something unbelievably special about a decillion (which we can write more succinctly as 10^{33}). Before revealing the strange answer, let us get an idea about how large 1 decillion is. It is greater than the number of atoms in a human breath (10^{21}). However, it's smaller than the number of electrons, protons, and neutrons in the universe (10^{79}).

What I find most interesting is that 10^{33} is the largest power of 10 known to humans that can be represented as the product of two numbers that themselves contain no zero digits:

$$10^{33} = 2^{33} \times 5^{33} = (8,589,934,592) \times (116,415,321,826,934,814,453,125)$$

For a variety of technical reasons, some mathematicians believe that no one will ever be able to find a larger power of 10 that can be represented as the product of two numbers that themselves contain no zero digits.

Do you think humanity will ever find such a number? My opinion is that the answer is "no."

Scott Bales from North Carolina notes that any possible solution must be of the form $2^x \times 5^x = 10^x$. If this is not true, one of the multiplicands' terms will have both 2 and 5 as factors, and the last digit of this term will be zero. The problem therefore is to find a power of 2 and a power of 5 that do not have zeros in them. Scott has written a Turbo Pascal program (running on a 486DX) to check 5^x for all values of x less than 60,000 (see program code in the next section). The only power of 5 that his program found that was greater than 5^{33} and also contained no zeros was 5^{58}. However, the power of 2 for $x = 58$ yielded a number with at least one zero.

Scott says, "Do I think such a number exists? I don't know—early evidence doesn't look good. If it exists, do I think humanity will find it? Yes."

Try It Yourself

Listed below is the source code Scott Bales used to check 2^x for huge x (up to 425,000). He could not find a value of x greater than 100 for which 2^x contained no zeros. This search required almost 30 hours on a 486DX33. The program is written in Turbo Pascal 3.0 (Borland).

```
program search;
const
  multiplier = 2;     /* This is the number we are checking powers of */
  maxsize = 32000;    /* This is how big our array can get */
  base = 10000;       /* This is the max size of each element of the
                         array */

var
  number : array [0..maxsize] of integer;
                      /* This is the array that holds the current
                         result */
  temp,               /* Temporary variable, used in zero
                         determination */
  loop1,              /* Loop variable, used in for loop */
  loop2,              /* Loop variable, used in for loop */
  result,             /* Used in array multiplication process */
  carry,              /* When one array element "overflows" into the
                         next */
  numtop,             /* Points to current top of the array */
  power2 : integer;   /* Counts iterations between updates */
  endflag,            /* Set just before we blow the top of the array */
  winner : boolean;   /* Set if the current number contains no zeros */
  power : real;       /* Current power we are checking - must be real,
                         as we're checking well beyond maxint */

begin

/* Initialize */

  number[0] := multiplier;
  power := 1;
  power2 := power2;
  numtop := 0;
  carry := 0;
  endflag := false;
  writeln('Powers of ', multiplier, ' that don't contain a 0:');
  writeln('1');

/* Main loop - continue while not overflow condition */

  repeat
    power := power + 1;
    power2 := power2 + 1;

/* Write out progress report every 10,000 iterations */

    if power2 >= 10000 then begin
      power2 := power2 - 10000;
      writeln('Currently processing 2 to the ', power)
    end;
```

```
/* Find the next power */

   for loop1 := 0 to numtop do begin
     result := (multiplier * number[loop1]) + carry;
     number[loop1] := result mod base;
     carry := result div base
   end;
   if (carry > 0) then begin
     if (numtop = maxsize - 1) then endflag := true;
     numtop := numtop + 1;
     number[numtop] := carry;
     carry := 0
   end;
```

```
/* Check to see if this is a winner - i.e., no zeros in the
   expansion */

   winner := true;
   for loop2 := 0 to numtop do begin
     temp := number[loop2];
     winner := (winner and ((temp mod 10) > 0));
     if (temp > 9) or (loop2 <> numtop) then begin
       temp := temp div 10;
       winner := (winner and ((temp mod 10) > 0));
       if (temp > 9) or (loop2 <> numtop) then begin
         temp := temp div 10;
         winner := (winner and ((temp mod 10) > 0));
         if (loop2 <> numtop) then
           winner := (winner and ((temp div 10) > 0))
       end
     end
   end;
   if winner then writeln(power)
```

```
/* If we haven't run out of room in the array, do it again */

   until endflag;
```

```
/* How far did we get? */

   writeln('Last power checked:  ', power)
end.
```

C H A P T E R 17

The history of proportion in art and architecture has been a search for the key to beauty.
— Jay Kappraff

Sketching with her slender pointed foot, some figure like a wizard pentagram, on garden gravel . . . .
— Alfred, Lord Tennyson

Infinite Star Chambers

Since prehistoric times, human emotions have been dominated by both fear and hope with respect to the invisible powers behind life's unpredictable events. Often humans have attributed these powers to demons or to good spirits, and interestingly, humans have invented various geometrical symbols to fight or appease such impalpable forces. One such magical symbol is the five-pointed star (Figure 17.1), or pentagram, used in various forms of witchcraft, magic, religion, and mysticism. The pentagram was even sometimes carved on babies' cribs or animals' stalls to protect against evil.

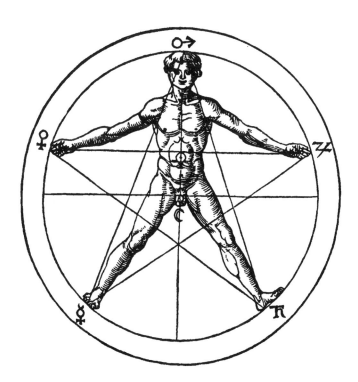

FIGURE 17.1. Symbolic representation of man as microcosmos (Agrippa).

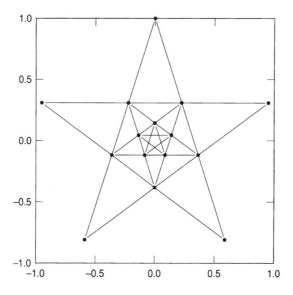

FIGURE 17.2.   Stars within stars.

As you may know, five-sided stars produced by drawing a continuous line with your pencil can nest inside one another (Figure 17.2). For example, a smaller star can sit inside the pentagon formed by the intersecting lines of a larger star. If you repeat the process over and over, enclosing one start within another, attractive nested stars result. (I used the set of C program codes given in Appendix 1 for this chapter to draw Figure 17.2.)

Historically, this pentagram (also known as the pentalpha, pentacle, or pentangle), formed by connecting nonconsecutive vertices of a regular pentagon, played many important roles. For example, it was the chosen symbol and seal of the Order of Pythagoreans, a secret society of ancient Greek mathematicians in the sixth century B.C.E. These Greeks also believed that nature could be explained by numerical relationships.

With this brief introduction to nested five-pointed stars, it is now time to boggle the mind. To do some quick experiments, draw a five-sided star formed with five line segments, each segment 1 inch long. Draw a larger star that encompasses this original star as suggested in Figure 17.2. We will call this star 2. As you continually nest stars, creating star 3, star 4, and so on, the assembly of stars grows larger at a fairly rapid rate.

I asked colleagues two questions, and most were astounded by the answers:

1. How many nestings N are required to make star N have an edge length at least equal to the diameter of the sun (4.567×10^9 feet, or 5.48×10^{10} inches).
2. How many nestings N are required to make the *cumulative* length of all the lines of all the nested stars equal to the diameter of the sun?

Take a quick guess before reading further. Do you think 100 nested stars would produce a five-pointed star as large as our sun, or would it take 1,000, or 10,000?

To answer this question, we need the value for the ratio of sizes between successive stars. It turns out that the ratio between side lengths of successive pentagrams is

$$r = (3 + \sqrt{5})/2 = 1 + \varphi = 2.618 \ldots$$

where φ is the famous golden ratio. To understand this, it helps to draw a "chord" connecting two nonconsecutive vertices of a regular polygon. The ratio of the chord's length to the length of one of the pentagon's sides is the golden ratio,

$$(1 + \sqrt{5})/2.$$

From this, it is possible to derive the length of the "chord" (the segment length) of the next bigger star compared to the length of the "chord" of its inscribed star. This ratio is the square of the golden ratio, which, interestingly, is the same value as the golden ratio plus 1.

Using this ratio, one can determine that the smallest N for which $r^N >$ 5.48×10^{10} inches is $N = 26$, with $r^{26} = 7.37\times10^{10}$. Therefore, it requires a mere 26 nested stars to make a final star slightly bigger than our own star, the sun! After only 7 nestings, the star is a big as a brontosaurus. In 15 nestings, it is as long as the Panama Canal. In 65 nestings, it will fill our entire universe. (See Table 17.1 for the growth rate.)

Next consider the second question about how many nestings N are required to make the *cumulative* length of all the lines of all the nested stars equal to the diameter of the sun. (Many colleagues thought the answer to this question would be far smaller than for the first question.) To solve this, one first notes that this is the sum of a geometric sequence with the ratio being the golden ratio squared (or the golden ratio plus 1). (A geometric sequence is a sequence for which the ratio of a term to its predecessor is the same for all terms, such as $s, sr, sr^2, sr^3, \ldots$.) Therefore, in our case,

$$s_0 = 1 \text{ and } s_n = s_{n-1}((1 + \sqrt{5})/2)^2.$$

The smallest N for which

$$5\times[r^{(N+1)} - 1]/(r - 1) > 5.48\times10^{10}$$

is 24, with

$$5\times(r^{25} - 1)/(r - 1) = 8.70\times10^{10}$$

(The formula gives the sum of the geometric series. The 5 value is used because each star had five edges, which we need to include in the sum.) Therefore, we would need the cumulative edges of 24 nested stars to form a line as big as the diameter of the

TABLE 17.1.   Star Chamber Growth

N^*	Edge	Length (inches)
0	1	
1	2.61803	
2	6.8541	
3	17.9443	
4	46.9787	
5	122.992	Height of large African elephant at shoulder
6	321.997	
7	842.999	Length of brontosaurus
8	2207	Height of Statue of Liberty
9	5778	Height of Washington Monument
10	15127	Length of *S.S. Queen Mary*
11	39603	Length of channel span, Golden Gate Suspension Bridge
12	103682	
13	271443	Height of Mt. McKinley
14	710647	
15	1.86050e+06	Length of Panama Canal
16	4.87084e+06	Length of Suez Canal
17	1.27520e+07	
18	3.33853e+07	
19	8.74037e+07	
20	2.28826e+08	Air distance: New York City to Los Angeles
21	5.99074e+08	Diameter of Earth
22	1.56840e+09	Circumference of Earth
23	4.10611e+09	
24	1.07499e+10	Mean distance: Earth to moon
25	2.81437e+10	Diameter of sun
26	7.36812e+10	
27	1.92900e+11	
28	5.05019e+11	
29	1.32216e+12	
30	3.46145e+12	Mean distance: Earth to sun
31	9.06219e+12	
32	2.37251e+13	Length of longest observed comet

(continued)

TABLE 17.1.   (continued)

N*	Edge	Length (inches)
33	6.21132e+13	Diameter of supergiant stars
34	1.62614e+14	
35	4.25730e+14	
36	1.11458e+15	
37	2.91800e+15	
38	7.63941e+15	
39	2.00002e+16	
40	5.23613e+16	
41	1.37084e+17	
42	3.58890e+17	One light year
43	9.39586e+17	
44	2.45987e+18	Distance to nearest star beyond the sun
45	6.44001e+18	
46	1.68602e+19	
47	4.41405e+19	Distance of 20 brightest stars in sky
48	1.15561e+20	
49	3.02544e+20	
50	7.92069e+20	
51	2.07366e+21	
52	5.42892e+21	
53	1.42131e+22	Distance of sun to center of galaxy
54	3.72104e+22	Diameter of Milky Way
55	9.74181e+22	
56	2.55044e+23	Distances of extragalactic nebulae
57	6.67713e+23	
58	1.74810e+24	
59	4.57657e+24	
60	1.19816e+25	
61	3.13683e+25	
62	8.21233e+25	
63	2.15002e+26	
64	5.62881e+26	
65	1.47364e+27	Radius of universe according to Einstein's cosmology

*N is the number of star nestings.

FIGURE 17.3.   Twelve-sided star chamber. This strange object seems to be growing at a slower rate than the star chamber in Figure 17.2. How many stars are required to make an edge of this object as long as the diameter of our sun?

sun. This result is only slightly less than the number of nestings needed to generate a single star's edge that is as big as the diameter of the sun!

A final problem to ponder is posed in Figure 17.3.

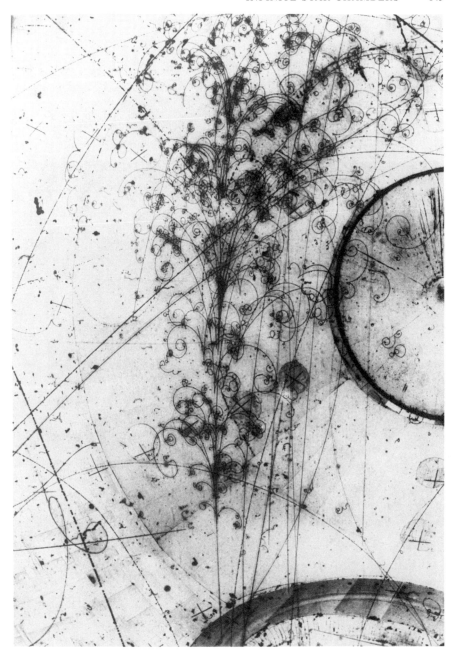

In this chapter we have been discussing unimaginably large objects. In contrast, how many smaller nestings are required to make the star as small as some of the subatomic particles that physicists have detected by observing trails in cloud and bubble chambers (above)? (For example, the radius of a hydrogen atom is 1.74×10^{-10} feet.) (Photo courtesy of NASA.)

Infinitely Exploding Circles

Math is a perfection in expression, like ballet or shaolin martial art.
— Venkata Guruprasad

Some scientists are bothered by infinities, which seem to crop up at embarrassing places in our theories of the universe. A black hole has an infinity at its very center.
— Fred Wolf, *Parallel Universes*

Draw a circle, with a radius equal to 1 inch. Next, circumscribe (that is, surround) the circle with an equilateral triangle. Next, circumscribe the triangle with another circle. Then circumscribe this second circle with a square.

Continue with a third circle, circumscribing the square. Circumscribe the circle with a regular pentagon. Continue this procedure indefinitely, each time increasing the number of sides of the regular polygon by one. Every other shape used is a circle that grows continually as it encloses the assembly of predecessors (Figure 18.1). If you were to repeat this process, always adding larger circles at the rate of a circle a minute, how long would it take for the largest circle to have a radius equal to the radius of our universe, 10^{26} feet?

By continually surrounding the shapes with circles, it would seem that the radii should grow larger and larger, becoming infinite as we continue the process. With a simple computer program, you can in fact show that the radius of a circle is always larger than its predecessor. (After all, the predecessor shapes are *enclosed* by the most recently added circles.) How much time is required for the circles to grow larger than our universe? The answer is that the assembly of nested polygons and circles will never grow as large as the universe, never grow as large as the Earth, never grow even as large as a basketball.

Although the circles initially grow very quickly in size, the rate of growth gradually slows, and the radii of the resulting circles approach a limiting value given by the infinite product:

$$R = 1/[(\cos\pi/3) \times (\cos\pi/4) \times (\cos\pi/5) \ldots] \qquad [1]$$

FIGURE 18.1. Exploding circles and polygons.

This can be written more succinctly with the appropriately impressive mathematical symbols:

$$R = \prod_{n=3}^{\infty} \left(\cos \frac{\pi}{n} \right)^{-1} \tag{2}$$

To better understand the process, you might like to think of the circle-growth process in terms of the circular cross-section of a balloon. An infinitely patient balloon blower gradually loses his wind as the balloons get ever larger, while approaching a limit radius R. Figure 18.2 shows circles with a radius determined by the previous infinite product formulas. To produce the figure, I plotted the circles for $n = 3$, $6, 9, \ldots, 99$.

This in itself is fascinating; however, this little assembly of circles and polygons has many other mind-numbing properties. Perhaps most intriguing is the controversy over the limiting value of R. It seems simple enough to compute. In fact, I have included programs in Appendix 1 to do just that. According to Kasner and Newman[1], who first reported a value in the 1940s, R is approximately equal to 12. A value of 12 is also mentioned in a German article published in 1964[2].

Now the fun begins. Hearing of my interest in unusual mathematical constants, C. J. Bouwkamp sent me his intriguing paper, published in 1965, that reports the true value of R ($R = 8.7000$).[3] I find it fascinating that, until 1965, mathematicians still assumed that the correct value of R was 12.

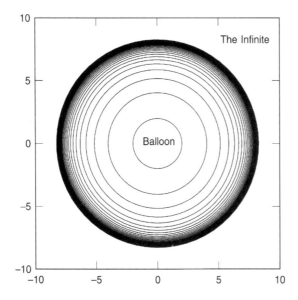

FIGURE 18.2.   An infinitely patient balloon blower gradually loses his wind as the balloon gets ever larger, while approaching a limit radius R.

Does this limiting radius (8.7000) really terminate so nicely in a continual stream of zeros? Alas, the answer turns out to be no. The correct value of R to 16 digits is:

$$R = 8.7000366252081945\ldots$$

As far as I know, to this day no one has computed a more accurate value for this infinite product.

What makes the problem even more fascinating is that it may be difficult for you to arrive at this value, even with a modern PC. The reason is that the infinite product in equation 1 or 2 is very slow to converge. Here are some values I obtained using the included program codes in Appendix 1 for a few sample iterations.

Iteration	Radius
100	8.28314
1,000	8.657231
10,000	8.695745

Although the value is getting closer to the reported accurate value of 8.7000 . . . , it is approaching awfully slowly. A related problem is posed in Figure 18.3.

FIGURE 18.3.   Exploding hexagons, an example of a nested set of polygons that diverges to infinity. Do nested hexagons explode faster than nested pentagons?

Not for the Meek

We are in the position of a little child entering a huge library whose walls are covered to the ceiling with books in many different tongues. . . . The child does not understand the languages in which they are written. He notes a definite plan in the arrangment of books, a mysterious order which he does not comprehend, but only dimly suspects. — *Albert Einstein*

For those of you who would not run from monstrous looking formulas, C. J. Bouwkamp has come up with the following esoteric transformation of the infinite product, which gives a very accurate value of *R* with far fewer iterations—because the reformulated infinite product converges much more quickly. For example, by using only 100 terms of the Bouwkamp formula, I find *R* is approximately equal to 8.700035. I wonder how many iterations would be required to achieve a similarly accurate value for *R* using the simpler iteration in the previous section (method 1)?

To compute approximations to *R* using the superaccurate method 2, use:

$$A = \frac{(1 - \pi^2/2 + `\pi^4/24) \times (1 - \pi^2/8 + `\pi^4/384) \times \pi^4\!\!\sqrt{24}}{\sin(\pi^2/\sqrt{6 + 2 \times \sqrt{3}}\,) \times \sin(\pi^2/\sqrt{6 - 2 \times \sqrt{3}}\,)}$$

$$B = \prod_{n=3}^{\infty} \frac{1 - \pi^2/(2 \times n^2) + \pi^4/(24 \times n^4)}{\cos(\pi/n)}$$

$$R = AB_n$$

Courageous readers may consult Bouwkamp's article for a derivation.[4]

CHAPTER 19

It is known that there are an infinite number of worlds, simply because there is an infinite amount of space for them to be in. However, not every one of them is inhabited. Therefore, there must be a finite number of inhabited worlds. Any finite number divided by infinity is as near to nothing as makes no odds, so the average population of all planets in the Universe can be said to be zero. From this it follows that the population of the whole Universe is also zero, and that any people you may meet from time to time are merely the products of a deranged imagination.

    — Douglas Adams, *The Restaurant at the End of the Universe*

The will is infinite and the execution confined. The desire is boundless and the act a slave to limit.

    — Shakespeare, *Troilus and Cressida*

The Infinity Worms of Callisto

If God were infinity, then divergent series would be God's angels flying higher and higher to reach Him or Her. Some angels fly faster than others, but given an eternity, all will return to their Creator in the Celestial Palace.

Let me explain with an example. Consider the following infinite series:

$$1 + 2 + 3 + 4 + \ldots .$$

If we add one term of the series each year, in four years the sum will be 10. Eventually, after an infinite number of years, the sum will reach infinity. Mathematicians call such series divergent series because they explode to infinity given enough terms. In the following tale, we are interested in series that diverge much more slowly than the previous example. We are interested in more magical series—angels, perhaps, with weaker wings.

On Callisto, a cold, icy moon of Jupiter, lives a race of intelligent worms called infinity worms. The number of years they are permitted to live is based on mathematical formulas that they are free to select early in their lives. (The government uses this approach to prevent overpopulation.)

The worms are particularly fascinated with divergent series that are simple to express. For example,

$$1 + \frac{1}{2} + \frac{1}{3} + \frac{1}{4} + \ldots$$

is an infinite series that grows in size as more and more terms are added. Of course, it explodes more slowly than $1 + 2 + 3 + \ldots$, but it does grow to infinity. If desired, the series may be written compactly as

$$\sum_{n=1}^{n=\infty} \frac{1}{n}$$

153

Some exploding series produce larger and larger values quickly, with just a few terms. Others grow very slowly, but *all* divergent series eventually reach infinity as more terms are added.

Once an infinity worm selects a particular series, it is permitted to live a number of years equal to the number of terms required to make the sum greater than 10. In other words, a very slowly growing series is desirable because it allows a worm to live many years. It turns out, as we shall soon see, that it takes a very large number of terms before

$$1 + \frac{1}{2} + \frac{1}{3} + \frac{1}{4} + \ldots$$

gets as large as 10. Naturally, the desire to discover slowly growing series gives the worm citizenry an incentive to become proficient at mathematics, and therefore a highly intelligent society has evolved over the years.

On one particularly frigid day, six colorful worms went to a burrow to meet their executioner.[1] The executioner worm stared down at a green worm. "What equation have you selected?" the executioner asked. The green worm responded with some trepidation in his voice:

$$\frac{1}{\mathrm{loglog}3} + \frac{1}{\mathrm{loglog}4} + \ldots \qquad \text{Green Worm}$$

"This is indeed a very slowly growing series," said the executioner. "It will eventually reach infinity, but—"

The green worm interrupted. "Sir, I have computed that the sum is not greater than 100 until 116 terms are used."

"I am sorry," the executioner said. "We are interested in the number of terms required to make the sum greater than 10. You will therefore live one more year." The executioner etched the following table in the icy soil to show the green worm the number of terms required to make the sum greater than 3, 5, 10, 100, 1,000, and 1 million.

Series	Number of Terms Required to Make the Sum Greater Than:					
	3	5	10	100	1,000	1,000,000
Green	1	1	1	116	1,800	2.6E6

(The E in 2.6E6 is simply a convenient way of indicating 2.6×10^6.)

Next, a blue worm came up to the executioner and presented her equation:

$$\frac{1}{\mathrm{log}2} + \frac{1}{\mathrm{log}3} + \ldots \qquad \text{Blue Worm}$$

"You are doing better than your slower brother," the executioner said upon seeing the blue worm's series. "You will live 20 more years":

Series	Number of Terms Required to Make the Sum Greater Than:					
	3	5	10	100	1,000	1,000,000
Green	1	1	1	116	1,800	2.6E6
Blue	3	7	20	440	7,600	1.5E7

A violet worm was next, with:

$$1 + \frac{1}{\sqrt{2}} + \frac{1}{\sqrt{3}} + \ldots \qquad \text{Violet Worm}$$

The executioner sighed. "That is a *little* better." He turned to the violet worm. "You will live 33 more years. Don't any of you wish to strive for immortality?"

Series	Number of Terms Required to Make the Sum Greater Than:					
	3	5	10	100	1,000	1,000,000
Green	1	1	1	116	1,800	2.6E6
Blue	3	7	20	440	7,600	1.5E7
Violet	5	10	33	2,500	2.5E6	2.5E11

A bright crimson worm puffed up his chest and grinned. His equation was:

$$1 + \frac{1}{2} + \frac{1}{3} + \ldots \qquad \text{Crimson Worm}$$

The executioner open his eyes wider. "Say, you truly are a great worm. You will live for 12,390 more years."

Series	Number of Terms Required to Make the Sum Greater Than:					
	3	5	10	100	1,000	1,000,000
Green	1	1	1	116	1,800	2.6E6
Blue	3	7	20	440	7,600	1.5E7
Violet	5	10	33	2,500	2.5E6	2.5E11
Crimson	11	82	12,390	10E43	10E43,000	10E43,000,000

"That is nothing!" cried a worm as black as the night. He pushed aside the crimson worm and screamed out:

$$\frac{1}{2\log 2} + \frac{1}{3\log 3} + \ldots \qquad \text{Black Worm}$$

The executioner started gagging as he simultaneously tried to take a deep breath. "*Mon Dieu!*" he yelled as his tail twitched and his five hearts began beating at twice their normal rate. Even the green and blue worms began to cheer. "You will live," said the executioner, "another $10^{4,300}$ years!"

Series	Number of Terms Required to Make the Sum Greater Than:					
	3	5	10	100	1,000	1,000,000
Green	1	1	1	116	1,800	2.6E6
Blue	3	7	20	440	7,600	1.5E7
Violet	5	10	33	2,500	2.5E6	2.5E11
Crimson	11	82	12,390	10E43	10E43,000	10E43,000,000
Black	8,690	1.3E29	10E4,300	10E(5E42)		

A hush fell over the burrow, and a few curious worms from adjacent tunnels came closer to hear what the last worm in the group would say.

A small, transparent worm with emerald internal organs and infinity in his eyes came up to the executioner. He slowly scanned his friends in the burrow with his clear eyes, and then he spoke in a quiet voice:

$$\frac{1}{3\log3\log\log3} + \frac{1}{4\log4\log\log4} + \cdots \qquad \text{Transparent Worm}$$

The black worm started to gag. "That's impossible—" he shouted, but his cries were drowned out by the merry screams of others with wide grins on their lumbricoid faces. Some cheered madly. Others exuberantly slapped the icy ground with their tails.

The infinity worms' screams of jubilant exaltation rose to a fever pitch. The executioner trembled, and said, "Y-your sequence does not reach 10 for—" He paused, trying to gather his thoughts. "You will live for $10^{10^{100}}$ years, a googolplex years."

Series	Number of Terms Required to Make the Sum Greater Than:					
	3	5	10	100	1,000	1,000,000
Green	1	1	1	116	1,800	2.6E6
Blue	3	7	20	440	7,600	1.5E7
Violet	5	10	33	2,500	2.5E6	2.5E11
Crimson	11	82	12,390	10E43	10E43,000	10E43,000,000
Black	8,690	1.3E29	10E4,300	10E(5E42)		
Transparent	1	60–70	10E(10E100)			

After hearing the executioner's pronouncement of the transparent worm's near-immortality, the infinity worms lifted the transparent worm up and paraded him around Callisto. There were bands playing. An icy equivalent of champagne was poured in every household, and the worm king declared a national holiday.

Digression

To protect my sanity, I had to believe that infinity was a comprehensible thing. — Clarke Punkford

Arlin Anderson from Huntsville, Alabama, is an infinity-worm aficionado. (He notes that I should be careful to explain that "log" denotes the natural log, sometimes written as "ln.") Arlin did various experiments with the crimson-worm formula to find values more accurate than the ones I reported. Because he obtained different values using different precision arithmetic in a computer program, he was careful to use high precision. The series to term 12,366 has a value of 9.999962147, and 1/12,367 makes it 10.00004300. Therefore, the crimson worm lives exactly 12,367 years. Note that if the crimson worm were to have dropped the first term in the series, making it (1/2 + 1/3 + . . .), he would live almost three times as long! This series to term 33,616 is 9.9999879, and 1/33,617 makes it 10.000017.

Lorenz Beyeler, from IBM Switzerland, wrote to me with other worm suggestions. He suggested I add the following:

A gray worm emerges from the crowd, bows in front of the executioner, and says: "Sir, I have found a way to live forever:"

$$1 + \frac{1}{10^1} + \frac{1}{10^2} + \frac{1}{10^3} + \ldots$$

(Is this a divergent series?) Lorenz also suggests a golden worm formula:

$$1 + \frac{1}{1^1} + \frac{1}{2^2} + \frac{1}{3^3} + \ldots$$

CHAPTER 20

Such as say that things infinite are past God's knowledge may just as well leap headlong into this pit of impiety, and say that God knows not all numbers . . . What madman would say so? . . . What are we mean wretches that dare presume to limit His knowledge?

— St. Augustine

The Undulation of the Monks

The number 69,696 is a remarkable number and certainly my favorite of all the integers. Aside from being almost exactly equal to the average velocity in miles per hour of the Earth in orbit, it is also the surface temperature of some of the hottest stars, in degrees Fahrenheit. More important are its fascinating mathematical properties.

It is written (right here) that a Tibetan monk once presented this number to a student and said: "What do you find significant about 69,696?"

The student thought for a few seconds, and replied, "That is too easy, Master. It is the largest undulating square known to humanity." The teacher pondered this answer, and himself started to undulate in a mixture of excitement and perhaps even terror.

To understand the monk's passionate response, we must digress to some simple mathematics. Undulating numbers are of the form abababababab . . . . For example, 171,717 and 28,282 are undulating numbers. A square number is of the form $y = x^2$. For example, 25 is a square number. So is 16. An undulating square is simply a square number that undulates.

When I conceived the idea of undulating squares a few years ago, it was not known whether any such numbers existed. It turns out that $69,696 = 264^2$ is indeed the largest undulating square known to humanity, and most mathematicians believe we will never find a larger one.

Dr. Noam D. Elkies from the Harvard Mathematics Department wrote to me about the probabilities of finding undulating squares. The chance that a "random" number around x is a perfect square is about $1/\sqrt{x}$. More generally, the probability is $x^{(-1+1/d)}$ for a perfect dth power. As there are (for any k) only 81 k-digit undulants, one would expect to find very few undulants that are also perfect powers, and none that is very large. Dr. Elkies believes that listing all cases may be impossible using present-day methods for treating exponential Diophantine (integer) equations.

Bob Murphy used the software Maple V to search for undulating squares, and he discovered some computational tricks for speeding the search. He began by examining the last 4 digits of perfect squares (that is, he computed squares mod 10,000). Interestingly, he found that the only possible digit endings for squares that undulate are 0404, 1616, 2121, 2929, 3636, 6161, 6464, 6969, 8484, and 9696. By examining squares mod 100,000, then mod 1,000,000, then mod 10,000,000, and so on, he found that no perfect square ends in 40404, 6161616, 63636, 464646464, or 969696, thereby allowing him to speed further the search process. Searching all possible endings, he asserts that, if there is an undulating square, it must have more than 1,000 digits.

Dr. Helmut Richter from Germany is the world's most famous undulation hunter, and he has indicated to me that it is not necessary to restrict the "mod searches" to powers of ten, and that arbitrary primes work very well. He has searched for undulating squares with 1 million digits or fewer, using a Control Data Cyber 2000. No undulating squares greater than 69,696 have been found.

Other Undulants

Randy Tobias of the SAS Institute in North Carolina notes that there are larger undulating squares in other number bases. For example, $292^2 = 85,264 = 41,414$ base 12. And 121 is an undulating square in any base. (121 base n is $(n + 1)^2$).

Interestingly, we find that there are very few undulating powers of *any* kind in base 10. For example, a three-digit undulating cube is $7^3 = 343$. However, Randy Tobias conducted a search for other undulating powers and found that 343 is the *only* undulant he could find that is formed by raising a number to a power p. He has checked this for $3 \leq p \leq 31$ and for all undulants less than 10^{100}. Undulating powers are indeed rare!

Undulating prime numbers, on the other hand, are more common. For example, Randy has discovered the following huge undulating prime:

$$7 + 720 \times (100^{49} - 1)/99 =$$
72727272727272727272727272727272727272727272727272727272727272727272
72727272727272727272727272727272727272727

(It has 99 digits.) To find this monstrosity, he also used the software program called Maple. The program scanned numbers using two lines:

```
(a*10 + b)*(10**(2*(k+1)) - 1)/99
a + 10*(a*10 + b)*(10**(2*(k+1)) - 1)/99
```

for ($0 \leq k \leq 50$, $1 \leq a \leq n - 1$, and $0 \leq b \leq n - 1$). The Maple "isprime()" function was used to check whether a number is prime. Maple makes it possible to work with very large integers.

There are many other undulating primes with many digits. However, there does not seem to be any undulating prime with an even number of digits. (Considering that $ababab \ldots ab = ab \times 10101 \ldots 01$, we should not expect to find any even-digit undulating primes.) I am interested in hearing from readers who have searched for undulating primes with larger periods of undulation, such as found in the prime number 5,995,995,995 (which does not finish its last cycle of undulation).

Binary Undulants beyond Imagination

Finally, binary undulants are powers of 2 that alternate the adjacent digits 1 and 0 somewhere in their decimal expansion. For example, the "highest quality" binary undulant I have found is 2^{949}. It has the undulating binary sequence 101010 in it, which I have placed in parentheses in the following:

$2^{949} =$
47584541071289058009537999940796817924200326453100622689784699949
81(101010)29132939953445386063877003218873559161286175137614546 72
78574369826493065785952766280250550668943187159661659651146975 27
57984765426503524599059416795862009216282102716609115705865638 54
433745326052103604911620698931 2

Here 949 is called an undulation seed of order 6, because it gives rise to a six-digit undulation pattern of adjacent 1's and 0's. When I challenged mathematicians and programmers around the world to produce a higher-order binary undulant, many took up the challenge. The highest-quality binary undulant so far known to humanity was discovered by Arlin Anderson of Alabama. He was the first to find that $2^{1,802}$ contains an eight-digit binary undulation. After much hard work he also found that $2^{7,694,891}$ starts with the digits 10101010173 . . . , and a week later he discovered that $2^{1,748,219}$ gives rise to a ten-digit undulant! Because Arlin only checked the last 240 digits of each number, he feels it is almost certain that there is a bigger binary undulation somewhere in the first million powers of 2. Considering that $2^{1,000,000}$ contains around 300,000 digits, the chance of finding a 10101010101 or 01010101010 is large. (Arlin uses a custom C program for large integer computation. The program runs on an Intergraph 6040 UNIX workstation and on a 486 PC. Searching 240 digits in 2 million powers of 2 required 15 hours.)

How do binary undulants vary with the base b? For example, for the case of $b = 2$, there are many binary undulants. Is it possible that as b increases, the quality of the best known cases decreases?

The Fractal Golden Curlicue Is Cool

I soon saw, by the way in which the white man's track doubled and redoubled again, that the fellow could not be cutting such curlicues for nothing.

— C. F. Hoffman, *Greyslaer*

How the hell can a mathematical curve have a temperature?

— Anonymous graduate student

Curlicues are fantastic spirals or twists that seem to be everywhere we look. When we think of examples of curlicues, both the mundane and exotic easily come to mind, such as: the gentle curl of a fern tendril, the shape of an octopus's retracted arm, the death-form assumed by a centipede, the spiral intestine of a giraffe, the shape of a butterfly's tongue, the spiral cross-section of a scroll, the shape of the yellow brick road in Munchkinland in the film classic *The Wizard of Oz*, and even the characters of several written languages such as Farsi and Tamil.

In this chapter I am interested in the mathematical curlicues, and in particular will explore infinite, recursively spiralling patterns. This class of curlicues is quite fascinating to geometers not only because of their beauty and complexity but also because of their renormalization symmetries and various fascinating mathematical properties. ("For Further Reading," at the end of this book, offers some interesting technical papers for the mathematically inclined reader.)

There are many different ways of generating various curlicues in the mathematical literature. My favorite method, developed by John Sharp and others, is quite easy to implement. If you want to draw these curlicues, you start by placing a dot on a piece of paper. Next you draw a line at a particular angle to the horizontal. From your new point at the end of the line, you draw another line of the same length but at a new angle. The process is continued until an intricate curve is drawn. What happens if you were to draw an infinite number of little line segments? What would the resulting diagram look like?

Let me begin by describing the mathematical recipe. Readers not interested in implementing the formulas can still appreciate the figures derived from a very sim-

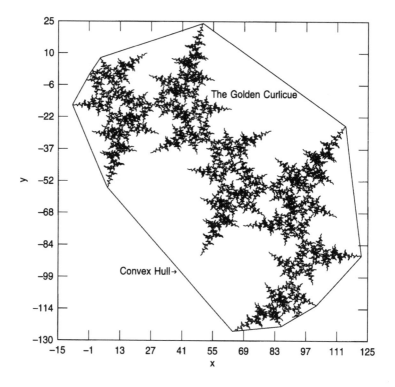

FIGURE 21.1.   The golden curlicue and its convex hull.

ple generating method. Let us first consider the angles and lines. The angle of each line segment is defined by multiplying a complete rotation (2π) by the constant s. To generate these curlicues with a computer program, the angle θ of each line segment in the plane is determined by $\theta_{n+1} = (\theta_n + s)\mathrm{mod}2\pi$ as n is incremented from 0 to 10,000, and s is a transcendental or irrational constant multiplied by 2π. The length of each line segment is constant. (θ is with respect to the previous line, not the horizontal.) The program code should clarify the algorithm. Generally, the resultant curves are fractal, exhibiting similar structures at different size scales.

A variety of curves can be generated with this approach for different constants, and the curves can be studied using thermodynamics (see examples in Figures 21.1 and 21.2). Traditionally, thermodynamics deals with the statistical properties of gases and involves quantities such as temperature, pressure, volume, and entropy. Who would guess that quantities such as temperature have recently been used by several authors to describe mathematical curves instead of gases?

I do not derive the following formula, because the references (see "For Further Reading") provide sufficient information for the mathematical reader. Of all the thermodynamic parameters one can study, I find a curve's temperature the most fascinating. The temperature T of curves Γ can be computed from

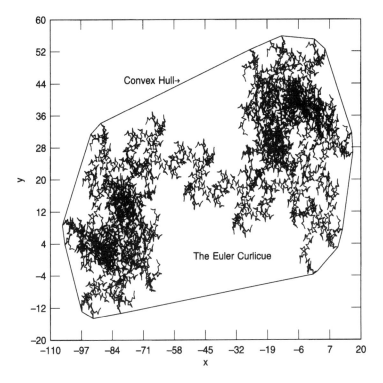

FIGURE 21.2.   The Euler curlicue and its convex hull.

$$T = \frac{1}{\ln[2l/(2l - h)]}$$

where l is the length of Γ and h is the length of a surrounding polygon known as its "convex hull." Specifically, the convex hull is the length of the perimeter of the smallest convex polygon that contains the curlicue. One easy way to understand a convex hull is to think of it as a rubber band placed around a set of points and contracted as far as possible. The temperature of a curve is zero if and only if the curve is a straight line. The hotter a curve becomes, the more wiggly it can be. (Similarly, the entropy, volume, and pressure of curves can be computed.)

I tested a range of famous transcendental and irrational constants that control the angle that each step takes in the curlicue. After 10,000 iterations (steps), the golden curlicue, defined by the golden ratio

$$s = (\sqrt{5} + 1)/2,$$

seems to have the lowest temperature. I am interested in hearing from readers who have tried other famous transcendental or irrational constants and found a cooler curlicue than the golden one. Below is a list of the temperatures measured at itera-

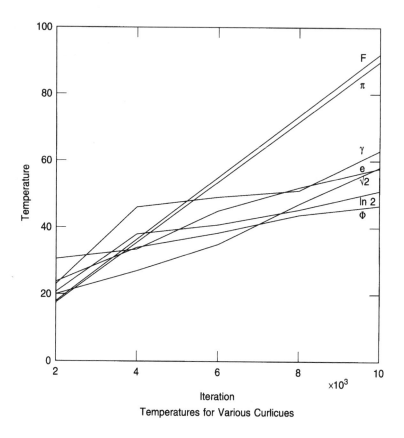

Temperatures for Various Curlicues

FIGURE 21.3. Temperature vs. iteration for various curlicues: F, the Feigenbaum number (4.669201); π; γ, Euler's constant (0.5772156); e; $\sqrt{2}$; ln2; and, φ, the golden ratio (1.618034). The temperature is sampled at 2,000, 4,000, 6,000, 8,000, and 10,000 iterations.

tion 10,000 for a number of important constants. Figure 21.3 shows the temperatures of various curlicues changing as they grow.

Curlicue Constant			**Temperature**
Golden ratio	=	1.618034	46 degrees
ln(2)	=	0.693147	51 degrees
e	=	2.718282	58 degrees
Sqrt(2)	=	1.414214	58 degrees
Euler's number	=	0.577215	63 degrees
pi	=	3.141592	90 degrees
Feigenbaum constant	=	4.669201	92 degrees

The fact that the golden curlicue (Figure 21.1) is quite cool will no doubt capture the imagination of aficionados of this constant—individuals often curious about the role the golden section (also called the golden mean or the divine proportion) plays in mathematics, art, architecture, literature, and esthetics. I thank Rajan Vadakkedath for assistance with hull computation and John Sharp for bringing fractal curlicues to my attention in a newsletter called *The Fractal Report*.

Curlicues and the Infinite

From a theoretical standpoint, the curves in this chapter are infinite. In other words, as they are generated by taking more and more steps on a plane, their lengths grows to infinity. I have computed the temperatures for the curves up to 10,000 steps. Would the temperature trends change if I were to use 1 million steps? Does the behavior at 10,000 steps reflect the limiting behavior near infinity? How does the finite precision of computers affect our ability to accurately draw curlicues?

Postscript

Mathematically inclined readers may be interested in other recursively spiralling patterns Γ created in the complex plane. One such curlicue can be generated using

$$\Gamma_L(\tau) \equiv \sum_{n=1}^{L} \exp(i\pi\tau n^2)$$

Here the terms (steps) are regarded as unit vectors, and L is made to increase indefinitely with the parameter τ held fixed. The resultant patterns are intricate superpositions of spiral structures whose scalings are sensitive to the value of τ.

More recently, experiments have been described that use a sequence of real numbers s_n that determine a unique direction in the plane, forming an angle $2\pi s_n$ radians with respect to the horizontal. The numbers s_n between 0 and 1 provide all possible directions so that direction depends only on the fractional part of s_n (that is, $s_n \bmod 1$). Graphically speaking, one draws a line of fixed length in the direction corresponding to s_0. With this new starting point, a line is drawn of the same length in a direction corresponding to s_1. In a computer program, this is implemented as $x_{n+1} = x_n + \cos(2\pi s_n)$, $y_{n+1} = y_n + \sin(2\pi s_n)$.

C H A P T E R 22

The brain is a three pound mass you can hold in your hand that can conceive of a universe a hundred-billion light-years across.

— Marian Diamond

Number is the bond of the eternal continuance of things.

— Plato

The Loneliness of the Factorions

Factorions are numbers that are the sum of the factorial values for each of their digits.[1] For example, 145 is a factorion because it can be expressed as

145 = 1! + 4! + 5!

Two tiny factorions are

1 = 1! and 2 = 2!

The largest known factorion is 40,585; it can be written as

40,585 = 4! + 0! + 5! + 8! + 5!

(The factorion 40,585 was discovered in 1964 by R. Dougherty using a computer search.) Can you end the loneliness of the factorions? Do any others exist?

Various proofs have been advanced indicating that 40,585 is the largest possible factorion, and that humans will never be able to find a greater factorion. In fact, these four factorions are the *only* factorions known to humanity. How can this be?

A more fruitful avenue of research may be the search for factorions "of the second kind," which are formed by the *product* of the factorial values for each of their digits. Additionally, hypothetical factorions "of the third kind" are formed by grouping digits. For example, a factorion of the third kind might have the form

abcdef = (*ab*)! + *c*! + *d*! + (*ef*)!

where each letter represents a digit. (Any groupings of digits are allowed for factorions of the third kind.) To date, I am unaware of the existence of factorions of the second or third kind, and I am interested in hearing from readers who can find any.[2]

Digression: Narcissistic Numbers

Dik T. Winter from Amsterdam is an expert on a somewhat related class of numbers that are the sums of powers of their digits. In other words, these *n*-digit numbers are equal to the sum of the *n*th powers of their digits. For example, $153 = 1^3 + 5^3 + 3^3$. Variously called narcissistic numbers, "numbers in love with themselves," Armstrong numbers, or perfect digital variants, these kinds of numbers have fascinated number theorists for decades. The English mathematician Godfrey Hardy (1877–1947) noted that, "There are just four numbers, after unity, which are the sums of the

TABLE 22.1.

Base	Total	Dig	Largest Narcissistic Number (number of digits)
2	1	(1)	1
3	5	(3)	122
4	11	(4)	3303
5	17	(14)	14421440424444
6	30	(18)	105144341423554535
7	59	(23)	12616604301406016036306
8	62	(29)	11254613377540170731271074472
9	58	(30)	104836124432728001478001038311
10	88	(39)	115132219018763992565095597973971522401
11	134	(45)	12344AA12A721803422912A8AA4963568083A268456A4
12	87	(51)	15079346A6B3B14BB56B395898B96629A8B01515344B4B0714B

cubes of their digits . . . . These are odd facts, very suitable for puzzle columns and likely to amuse amateurs, but there is nothing in them which appeals to the mathematician." I gave 153 as an example of such a number. Can you find the other three? (The second program for this chapter in Appendix 1 will help you search for these numbers.)

The largest narcissistic number discovered to date is the incredible 39-digit number

115132219018763992565095597973971522401

(Each digit is raised to the 39th power.) Can you beat this world record? What would Godfrey Hardy have thought of this multidigit monstrosity?

The frequency of occurrence of narcissistic numbers varies according to the base of the number system in which one conducts a search. For example, in our standard (base 10) system are 88 known numbers of this type, whereas in base 4 there are only 11. Table 22.1 lists numbers from Dik T. Winter. The number of digits for the largest narcissistic number known is in parentheses.

As one searches for larger and larger narcissistic numbers, will they eventually run out, as in the case of the lonely factorions? If they are proved to die out in one number system, does this mean they are finite in another? (Martin Gardner wrote to me recently and indicated that the number of narcissistic numbers has been proved finite. They cannot have more than 58 digits in our standard base 10 number system.)

Finally, Kevin S. Brown writes that he knows of only three occurrences of $n! + 1 = m^2$, namely $25 = 4! + 1 = 5^2$, $121 = 5! + 1 = 11^2$, and $5,041 = 7! + 1 = 71^2$. We do not know whether there are any others. Perhaps these "Brown numbers" will be as lonely as the factorions. The prolific mathematician Paul Erdös long ago conjectured that there are only three such numbers. Erdös offers a cash prize for a proof of this!

Proof

Several colleagues have pointed out that an n-digit factorion cannot exceed $n \times 9!$ (because each digit has a maximum value of 9). One can use this fact in conducting computer searches for factorions.

Matthew M. Conroy from Colorado has given me permission to include his proof that only four factorions of the first kind exist. The basic idea of his proof is as follows. Let $k(n)$ be the function that gives the sum of the factorial of each digit of a positive integer n (in decimal notation, base 10). For example, $k(32) = 3! + 2! = 6 + 2 = 8$, and $k(145) = 1! + 4! + 5! = 1 + 24 + 120 = 145$. For factorions, we suppose $k(n)=n$.

Next let d be the number of digits in integer n. Then $10^{(d-1)} \le n < 10^d$. (For example, if n has three digits, then $100 \le n \le 1,000$.) Now, because n has d digits and 9 is the largest possible digit, we can conclude that $k(n) \le 9! \times d$. Hence, $10^{(d-1)} \le n = k(n) \le 9! \times d$. In other words, $10 \le (9! \times d)^{(1/(d-1))}$. The right side is a decreasing function of d (for $d > 1$ at least); when $d = 7$, $(9!d)^{1/(d-1)}$ is approximately 11.68, and when $d = 8$ it is approximately 8.38. From this we can conclude that $d \le 7$. Thus we have $n = k(n) = 9! \times 7 = 2,540,160$. With a computer, we can check all positive integers less than 2,540,160 to see whether they have the property $k(n) = n$. If we do this, we find that only 1, 2, 145, and 40,585 have this property.

C H A P T E R 23

Beauty is the first test: there is no permanent place in the world for ugly mathematics.

—G. H. Hardy

Mathematics, rightly viewed, possess not only truth, but supreme beauty—a beauty cold and austere, like that of sculpture.

—Bertrand Russell

Escape from Fractalia

I first came across the beautiful graphic work of Dr. J. Clint Sprott when he submitted a paper titled "Automatic Generation of Strange Attractors" to *Computers and Graphics*, a journal for which I am the associate editor. I accepted his paper without hesitation and was delighted to see a more thorough exposition in his popular *Strange Attractors: Creating Patterns in Chaos* (1993). We soon grew to be good friends, often communicating on the Internet, and we naturally could not resist a collaboration dealing with our mutual interests. This chapter is a result of that collaboration. As I mention later, Dr. Sprott generated numbers on his personal computer in Wisconsin and electronically sent them to me in New York, at which point my programs read them and rendered them using a graphics supercomputer. (Do not let this scare you—the patterns are easily generated on a personal computer.) What you see here is also the result of our collaborative writing.

An infinite number of monkeys with an infinite number of typewriters (word processors, nowadays) will eventually reproduce every work of Shakespeare. The problem is that someone has to sort through all the gibberish to find the occasional gem. On the other hand, the criteria for visual art are much less constrained than for literary composition. A monkey with a paintbrush has a reasonable chance of producing a pattern that someone would consider artistic.

Dr. Sprott and I have been experimenting with replacing the monkey by a computer and having it generate a collection of visually interesting patterns.[1] Unlike the monkey, the computer can be trained to select those images that are likely to appeal to humans. The procedure is to take some simple equations with adjustable coefficients chosen at random, solve the equations with the computer, and display those that meet criteria that we have found to be strong indicators of artistic quality.

One approach is to use equations with chaotic solutions. Such solutions are unpredictable over the long term yet exhibit interesting structures as they move about on a strange attractor, a fractal object with noninteger dimension. Our books show many examples of computer art produced by these and related methods.[2,3] In this chapter, we propose another simple method for producing fractal art.

Simple Equations Make Interesting Patterns

Some people can read a musical score and in their minds hear the music.
. . . Others can see, in their mind's eye, great beauty and structure in

173

*certain mathematical functions. . . . Lesser folk, like me, need to hear
music played and see numbers rendered to appreciate their structures.
— Peter B. Schroeder*

Traditionally when physicists or mathematicians saw complicated results, they often
looked for complicated causes. In contrast, many of the shapes that follow describe
the fantastically complicated behavior of the simplest of formulas (Figures 23.1 and
23.2). The results should be of interest to artists and nonmathematicians, and anyone
with imagination and a little computer programming skill.

One lesson of chaos theory is that simple, nonlinear equations can have complicated solutions. The solutions are most interesting if they involve at least two variables, x and y, that can be used to represent horizontal and vertical positions, and if
the variables are advanced step by step in an iterative process. The simplest non-
linearities are quadratic (x^2 or xy). The most general two-dimensional quadratic iterated map is:

$$xnew = a + bx + cx^2 + dxy + ey + fy^2$$
$$ynew = g + hx + ix^2 + jxy + ky + ly^2$$

FIGURE 23.1.   Escape-time fractal for parameters EHPLWTDJRCAP. (See text for explanation of parameter code. Also see color plates 3–6 and 10–12 for additional examples.)

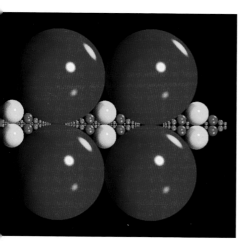

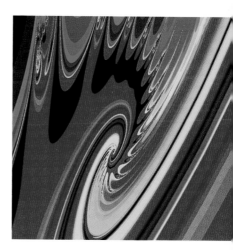

1

2

3

1   Three-dimensional representation of Ford froth. The color of the spheres is related to their radii. Transparency is used to visualize some of the internal Chrysler spheres. See "Fractal Milkshakes and Infinite Archery" (chapter 14) for more information.

2   Demagnification of color plate 1.

3   Escape-time fractal for parameters JLLGJTKDCRWY. See "Escape from Fractalia" (chapter 23) for more information.

4 Escape-time fractal for parameters EHPLWTDJRCAP. See "Escape from Fractalia" (chapter 23) for more information.

5 Mount Fractalia, a three-dimensional representation of the previous color plate.

6 The fractal curtain, an escape-time fractal for parameters WBMLNRQMNRAA. See "Escape from Fractalia" (chapter 23) for more information.

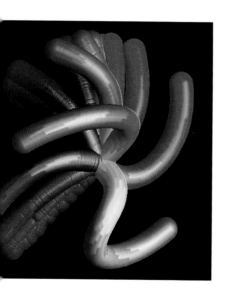

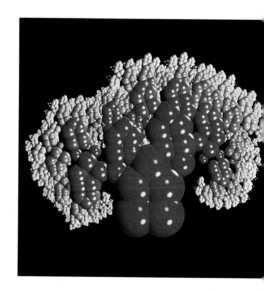

7 Artificial terrain synthesized using the logit transformation of the logistic variable. See "Computers, Randomness, Mind, and Infinity" (chapter 31) for more information.

8 Noise-sphere computed from a bad random number generator with correlations between digits. To give a better feel for the global structure produced, spheres are larger than used for the other noise-sphere figures and interpenetrate to a greater degree. See "Computers, Randomness, Mind, and Infinity" (chapter 31) for more information.

9 Coral-like form produced using recursive programming methods. See "Recursive Worlds" (chapter 27) for more information.

10   Escape-time fractal for parameters MWRQFVMKUVFK. See "Escape from Fractalia" (chapter 23) for more information.

11 Escape-time fractal for parameters KUONOVSVFLAR. See "Escape from Fractalia" (chapter 23) for more information.

12 Escape-time fractal for parameters EMXIXPQMMTWRJ. See "Escape from Fractalia" (chapter 23) for more information.

where a through l are constants chosen at random but held the same throughout the calculation. Think of this as a mathematical feedback loop in which new values for x and y are used again in the next round of iteration by setting $x = xnew$ and $y = ynew$. The constants can be considered as settings of a combination lock, each revealing a different pattern for you to admire. They open doorways to an infinite reservoir of magnificent shapes and forms.

There are many ways to display the solution to our general maps. One way is to plot successive values of x and y as dots on the screen. Many solutions will move toward a point and remain there. Others will settle into a periodic orbit or will move off toward infinity. The interesting cases are the chaotic ones that remain confined to a limited region but whose orbits produce a strange attractor with intricate fractal structure. You can choose the starting values of x and y arbitrarily within the basin of attraction, but you should discard the first few iterates because they probably lie off the attractor.

Another way you can display the solution is to solve the equations with many different starting values and count the number of iterations required for the solution to wander outside some region in the xy-plane. You can use the final number of iterations to determine the color for that point in the plane of initial xy-values. Plots produced in this way are called escape-time fractals because the color contours indicate the time required for the orbit to escape from the region. Some initial values may have orbits that never escape, and so you need to have a "bailout condition" beyond which your program stops iterating and moves on to test the next initial condition.

Some of the most artistic examples of escape-time fractals explored in the past have been Julia sets, fractal representations describing the behavior of the map $znew = z^2 + c$, where z and c are complex numbers.[4] In terms of x and y this map is

$$xnew = x^2 - y^2 + a$$
$$ynew = 2xy + b$$

and the test escape region is normally taken as a circle of radius of 2 centered on the origin. These traditional Julia sets are a special case of the more general maps we propose in this chapter.

Computer Art Critic

There is no excellent beauty that hath not some strangeness in the proportion. — Francis Bacon

Visually interesting escape-time fractals are those for which the orbits escape slowly. We have found that escape times between about 100 and 1,000 produce the most fascinating maps. We start by choosing the 12 coefficients randomly over the range -1.2 to 1.2 in increments of 0.1, and then we iterate the equations for our generalized quadratic map with initial conditions of $x = y = 0$. If we find a group of coefficients that result in the initial point escaping beyond a circle of radius 1,000 centered at the origin within 100 to 1,000 iterations, then we save the parameter set and compute an escape-time fractal for a particular region of the xy-plane. The choice of 1.2 for the

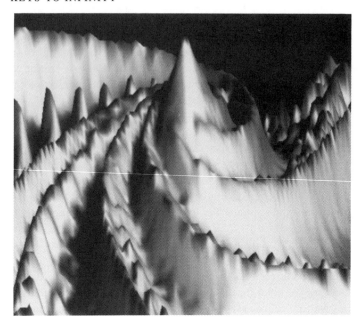

FIGURE 23.2.   Mount Fractalia, a 3-D representation of Figure 23.1.

coefficient range is about optimal for speeding the search, and it allows the coefficients to be compactly represented as letters of the alphabet (A = –1.2, B = –1.1, through Y = 1.2) for easy reference and replication.

The BASIC program in Appendix 1 produces an endless succession of escapetime fractals by this method. It should run without modification under QBASIC, QuickBASIC, VisualBASIC for MS-DOS, or PowerBASIC on the IBM PC. It assumes VGA graphics (640-by-480 pixels with 16 colors). When run on a 486DX33 under PowerBASIC 3.0, the program takes on average about 10 seconds to find each interesting case and about a minute to plot it. In the process of searching for what we call "beautiful parameters," it discards about 300 sets for every 1 it saves for plotting. This still leaves about 200 trillion cases, nearly all of which are different. If you view them at the rate of one per second, it would take more than 6 million years to see them all, and thus it is very unlikely that any of the patterns you produce will ever have been seen before. We often run the program overnight and capture the screens to graphics files, which can be rapidly examined the next morning.

Sample Artwork

The job of an artist is always to deepen the mystery. — Francis Bacon

Although the programs and methods we have described work well with personal computers and produce good-quality patterns within VGA screen limits, we became curious as to how our approach could be extended to higher resolutions. Naturally, high-resolution, high-iteration computations would take longer to perform. To overcome these problems and push our method to the limit, we made use of IBM's Power Visualization System, a graphics supercomputer that can compute our general quadratic iterated map in parallel using up to 32 Intel i860XP processors. This means the computer can simultaneously work on different portions of the artwork and therefore greatly reduces the time needed to explore a large number of images to find interesting structures. With our software and special hardware we can compute and view images at VGA resolution in less than a second. The images in this chapter and in some of the color plates for this book were computed at a resolution of 1,600-by-2,400 pixels using 1,000 iterations for each pixel. Computation and display required less than a minute for each high-resolution image. The final step in esthetic presentation involves the mapping of escape times to colors. Although beautiful images result from simple color mappings, such as the one used in the BASIC code in Appendix 1, we used custom software that permits the algorithmic or mouse selection of colors from a palate of 2^{24} colors in a convenient fashion. It is also possible for us to render the maps in three dimensions, representing escape time as height. The Power Visualization System enables us to "fly by" the 3-D maps in real time, permitting us to make videotape animations characterizing the intricate behavior of these functions.

In the future, high-end methods such as these will be accessible to everyone as PC power increases, cost decreases, and Internet collaborations such as ours become commonplace. In our own collaboration, input parameters were generated on a personal computer in Wisconsin, electronically sent over the Internet to New York, and then read by the Power Visualization System and rendered.

To artists and computer graphics programmers, the automated generation of our generalized quadratic-map representations has great appeal because the speed allows the human analyst the freedom to experiment with many parameters. Students and teachers will enjoy such an approach as they explore and demonstrate the complexity and chaos associated with simple formulas. Mathematicians may find the approach useful because the sheer number of different structures they can generate has the potential for making complex behaviors apparent that might have been overlooked using traditional approaches.

Further Suggestions

The method we have described is simple and effective, but we do not claim it is the final word on automated computer art. You can explore other criteria for selecting the patterns and other ways to display them. We chose a generalized two-dimensional quadratic map to emphasize the complexity and variety that arises from simple equations. You can easily extend the technique to other mathematical functions and to higher dimensions. The same ideas can be used for automatic generation and evaluation of computer music. We hope that we have introduced you to a new use for your computer, and we would like to hear of any interesting results you obtain.

C H A P T E R 24

When I consider the small span of my life absorbed in the eternity of all time, or the small part of space which I can touch or see engulfed by the infinite immensity of spaces that I know not and that know me not, I am frightened and astonished. . . .

— Blaise Pascal

Are Infinite Carotid-Kundalini Functions Fractal?

The humble sine waves that lie at the very foundation of trigonometry have a special beauty all their own. It takes just a little coddling to bring the beauty out. But who would guess, for example, that intricate fractal patterns lurk within the cosine operation applied to real numbers?

Consider the union of the infinite set of curves produced by Carotid-Kundalini (C-K) functions defined by:

$$y = \cos(n \times x \times \text{ acos } (x))$$

where $(-1 < x < 1, n = 1, 2, 3, \ldots)$, and acos designates the arccosine function. The set of superimposed curves is very simple to plot—most computer hobbyists could easily program and plot them on a personal computer—but the curves have an extremely complicated and beautiful structure (Figure 24.1). For example, for $x < 0$ there appears to be an exotic fractal structure with gaps repeated at different size scales and with progressively increasing spacing as x becomes smaller. This behavior is also illustrated in Figure 24.2, a magnification of Figure 24.1. For $x > 0$ there is an intricate oscillating web. Centered around $x = 0$ are a series of bell-shaped curves. You can compute the union of the first 25 curves using the following logic (a more complete program listing is provided in Appendix 1):

```
for (n=1; n<=25; n=n+1) {
    for (x=-1; x<=1; x=x+.01) {
        y = cos((float)n*x*acos(x));
        if (x==-1) MovePenTo(x,y);
            else DrawTo(x,y);
    }
}
```

If you have the ability to display these curves, make a plot from $-1 < x < 1$ and $-1 < y < 1$.

For the most graphical beauty, and for your eye to detect the most structure, scale the plot so the y-axis is much smaller than the x-axis, creating a thin strip. Try magnifying Fractal Valley (my personal favorite, to the left of zero), Oscillation Land (to the right of zero), and Gaussian Mountain Range (in the center). What happens as n approaches infinity? What do we know about the spacing of gaps in Fractal Land?

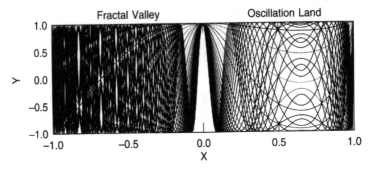

FIGURE 24.1. Carotid-Kundalini functions. Fractal Valley extends from $-1 < x < 0$. Oscillation Land extends from $0 < x < 1$. The term *Gaussian Mountain Range* refers to the central series of bell-shaped curves. The curves were plotted for $1 < n < 25$.

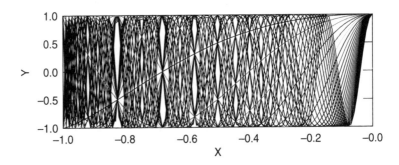

FIGURE 24.2. Magnification of left side of Figure 24.1.

Even though to my knowledge these curves have not been well characterized, we can be certain of several things. For example, they are bounded by $y = \pm 1$. Also, because acos(1) = 0 and cos(0) = 1, the infinite number of C-K curves contain the point (1, 1). This means that all curves must meet at the upper right section of Figure 24.1. It is almost as if some geometrical god has come down and placed a pin at (1, 1) to tie the majestic, unruly curves together.

The curves intersect the line $y = 1$ whenever any of the following conditions are met: $n = 0$, $x = 0$, or acos $(x) = 0$. This accounts for the tip of the bell-shaped curves in Gaussian Mountain Range. The bells are centered at the origin at $x = 0$. Zero crossings satisfy $\pi/2 = n \times$ acos (x).

The Magical Fractal Gaps

Dr. V. T. Rajan from IBM was the first person to quantify the exact placement of the mysterious repetitive gaps that occur in the plots for $-1 < x < 0$. The center of the gaps occurs whenever $x \arccos(x) = 2\pi(p/q)$ where p and q are integers and have no common factors. When these conditions are met then the C-K functions all assume only ceiling$((q+1)/2)$ distinct values. The term *ceiling* means to round the value up to the nearest integer. (Notice in Figure 24.2 only a few distinct values are assumed by the superimposed curves while in the gaps.) The distinct values that all the curves assume are $\cos(2\pi r/q)$, $r = 0, 1, \ldots$, floor $(q/2)$. The term *floor* means to round the value down to the nearest integer.

As more C-K functions are superimposed, the graph becomes denser (blacker). However, if we were to magnify the dense "fabric," we would see gaps satisfying the Rajan conditions. Can you write a graphics program permitting the search for gaps?

A Programming Tip for Some BASIC Users

Some forms of BASIC may not have an arccosine (ACOS) function, even though they all have an arctangent function, ATAN or ATN. Perhaps this omission has come about because arccosine (and arcsine) can be expressed in terms of arctan. In fact, some BASIC manuals give the explicit formulas for computing arccosine and arcsine.

William Rudge from California suggests that implementation of ACOS and ASIN may have been omitted to save memory, particularly in the past when computers had little memory. Perhaps certain trigonometric functions were considered to be used very seldom. Most other computer languages have the ACOS function.

To compute ACOS from ATN, one can use:

```
ACOS(y) = ATN(SQRT(1 - y*y)/y)
```

If y is close to zero, numerical errors will be large, so it may be useful to check the values of y, as in the QuickBASIC program provided by William Rudge in Appendix 1.

Details

Mathematically inclined readers may be interested in the work of Robert Israel from the Department of Mathematics at the University of British Columbia. Others will want to skip this section. To better understand the C-K functions, Robert finds it useful to create a new variable, t. Let $t(x) = x \times$ acos (x), and study the graph of t as a function of x. There is one maximum at (according to Maple Software) $x = .65218$ and $t = .56109$. Moreover, $t(-1) = -\pi$ and $t(1) = 0$. We have $y = 0$ when $n \times t$ is an odd multiple of $\pi/2$. There are $n - 1$ odd integers between $-2n$ and 0, and floor $(.17860 \times n + .5)$ between 0 and $2n \times .56109/\pi$ (where floor(x) = greatest integer $\leq x$). Therefore, the function crosses the y-axis $n - 1$ times for $-1 < x < 0$, and floor $(0.17860 \times n + .5)$ times for $0 < x < 0.65218$, and another floor $(0.17860 \times n + .5)$ times for $0.65218 < x < 1$. Robert assumes that 0.561096 is actually an irrational multiple of π, which is almost certainly true.

There will be a maximum whenever $\cos(nt) = 1$, that is, when nt is a multiple of 2π. There will be a minimum whenever $\cos(nt) = -1$, that is, when nt is an odd multiple of π. Moreover, there will be a maximum or minimum at $x = 0.65218$ where t has its maximum.

C H A P T E R 25

The rationality of our universe is best suggested by the fact that we can discover more about it from any starting point, as if it were a fabric that will unravel from any thread.

— George Zebrowski, 1994

The Crying of Fractal Batrachion 1,489

Batrachions are a class of bizarre and infinite mathematical curves that hop like frogs from one "lilypad" to the next as they parade along the number system. These little-known curves derive their name from *batrachian*, which means "froglike." (To pronounce the word, note that the "ch" has a "k" sound.)

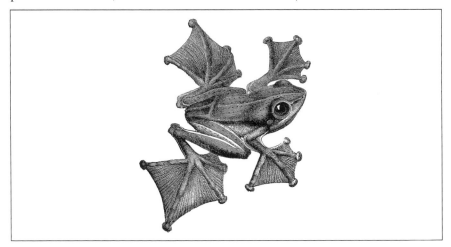

A batrachian preparing to hop.

In addition to hopping in a strange manner from integer to integer, batrachions also have other interesting properties. For example, they are often fractal, exhibiting an intricate self-similar structure when examined at different size scales. Also, they evolve from very simple-looking recursive formulas involving integers.

As background, perhaps the most common example of recursion in programming and in mathematics is one that defines the Fibonacci numbers. This sequence of numbers, called the Fibonacci sequence, plays important roles in mathematics and nature. These numbers are such that, after the first two, every number in the sequence equals the sum of the two previous numbers:

$$F_N = F_{N-1} + F_{N-2} \quad \text{for } N \geq 2, F_0 = F_1 = 1$$

This defines the sequence 1, 1, 2, 3, 5, 8, 13, 21, ...

With this brief background to recursion, consider my favorite batrachion, produced by the simple yet weird recursive formula

$$a(n) = a(a(n-1)) + a(n - a(n-1))$$

The formula for the batrachion is reminiscent of the Fibonacci formula in that each new value is a sum of two previous values—but not of the immediately previous two values. The sequence starts with $a(1) = 1$ and $a(2) = 1$. The "future" values at higher values of n depend on past values in intricate recursive ways. Can you determine the third member of the sequence? At first, this may seem a little complicated to evaluate by hand, but you can begin slowly by inserting values for n, as in the following:

$$a(3) = a(a(2)) + a(3 - a(2))$$
$$a(3) = a(1) + a(3 - 1)$$
$$a(3) = 1 + 1 = 2$$

Therefore, the third value of the sequence, $a(3)$, is 2.

The sequence $a(n)$ seems simple enough: 1, 1, 2, 2, 3, 4, 4, 4, 5, ... Try computing a few additional numbers. Can you find any interesting patterns? The prolific mathematician John H. Conway presented this recursive sequence at a talk he gave at AT&T Bell Labs entitled "Some Crazy Sequences."[1] He noticed that the value $a(n)/n$ approaches 1/2 as the sequence grows, and n becomes larger. Table 25.1 lists the first 32 terms of the batrachion and the ratio $a(n)/n$.

I first became interested in this sequence after reading Schroeder's delightful book *Fractals, Chaos, Power Laws*,[2] but, alas, there were no graphics included to help readers gain insight into the behavior of the batrachion. It turns out that this sequence has an incredible amount of hidden structure. Figure 25.1 is a plot of $a(n)/n$ for values of n between 0 and 200. Notice how the curve hops from one value of 0.5 to the next along very intricate paths. Each hump of the curve appears to be slightly

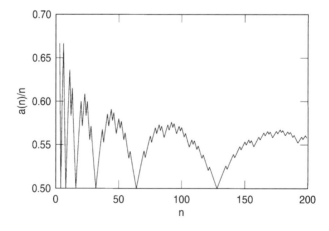

FIGURE 25.1. Batrachion $a(n)/n$ for $0 < n < 200$.

TABLE 25.1.   First 32 Terms of the Batrachion

n	$a(n)$	$a(n)/n$	n	$a(n)$	$a(n)/n$
1	1	1.0	17	9	.5294
2	1	1.0	18	10	.5555
3	2	.666	19	11	.5789
4	2	.5	20	12	.6
5	3	.6	21	12	.5714
6	4	.666	22	13	.5909
7	4	.5714	23	14	.6086
8	4	.5	24	14	.5833
9	5	.5555	25	15	.6
10	6	.6	26	15	.5769
11	7	.6363	27	15	.5555
12	7	.5833	28	16	.5714
13	8	.6153	29	16	.5517
14	8	.5714	30	16	.5333
15	8	.5333	31	16	.5161
16	8	.5	32	16	.5

lower than the previous, as though a virtual frog were tiring as it explored higher and higher numbers. As the frog nears infinity, will it stop its hopping and lie dormant at $a(n)/n = 0.5$? Figure 25.2 is a plot for the first 1,000 values of the sequence. Notice that it is very similar to Figure 25.1. More and more humps evolve with an intricate self-similar arrangement of tiny jiggles along the path.

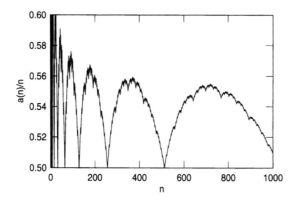

FIGURE 25.2.   Batrachion $a(n)/n$ for $0 < n < 1,000$.

$10,000 Cash Award

Let us consider how quickly the frog approaches its 0.5 destination at infinity. For example, can you find a value of n beyond which the value of $a(n)/n$ is so tiny that it is forever within 0.05 from the value 1/2? (In other words, $|a(n)/n - 1/2| < 0.05$. The bars indicate absolute value.)

A difficult problem? John Conway, the prolific British mathematician, offered $10,000 to the person to find the first value of n such that the frog's path is always less than 0.55 for higher values of n. A month after Conway made the offer, Colin Mallows of AT&T solved the $10,000 question: $n = 1,489.$[3] Figure 25.3 shows this value on a plot for $0 < n < 10,000$. (For a variety of minor technical reasons, a less accurate number is published in Schroeder's book.) As I write this, no one on the planet has found a value for the smallest n such that $a(n)/n$ is always within 0.001 of the value 1/2, that is $(|a(n)/n - 1/2| < 0.001)$. (No one even knows whether such a value exists.)

Looking at the curves, we can see the following. The frog hits the pond periodically. In fact, $a(n)/n$ "hits" 0.5 at values corresponding to powers of 2, such as at 2^k, $k = 1, 2, 3, \ldots$. Does each hump reach its maximum at a value of n halfway between the 2^k and 2^{k+1} endpoints?

Tal Kubo from the Mathematics Department at Harvard University is one of the world's leading experts on this batrachion. He notes that the sequence is subtly connected with a range of seemingly unrelated topics in mathematics: variants of Pascal's triangle, the Gaussian distribution, combinatorial operations on finite sets, and Catalan numbers. Tal Kubo and Ravi Vakil have developed algorithms to compute the behavior of the batrachion as it nears infinity. Indeed, they have found that the

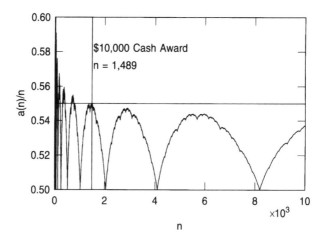

FIGURE 25.3.   Batrachion $a(n)/n$ for $0 < n < 10,000$.

TABLE 25.2.   The Infinite Frog

e	Last n Such That $|a(n)/n - 1/2| > e$
1/20	1,489 (found by Mallows in 1988)
1/30	758,765
1/40	6,083,008,742 (found by Mallows in 1988)
1/50	809,308,036,481,621
1/60	1,684,539,346,496,977,501,739
1/70	55,738,373,698,123,373,661,810,220,400
1/80	15,088,841,875,190,938,484,828,948,428,612,052,839
1/90	127,565,909,103,887,972,767,169,084,026,274,554,426,122,918,035
1/100	8,826,608,001,127,077,619,581,589,939,550,531,021,943,059,906,967,127,007,025

frog tires rather slowly! For example, the frog's jumps are not always less than 0.52 until it has jumped 809,308,036,481,621 times! Table 25.2 lists values for different frog jump heights. These values were found by Tal Kubo and Ravi Vakil using a Mathematica program running on a Sun 4 computer.

Colin Mallows, the statistician who conducted the first in-depth study of this class of curves, notes that no finite amount of computations will suffice to prove that the regularities we see in the curve persist indefinitely. He does note that the differences between successive values are either 0 or 1. Is this true indefinitely?

Visualizing Infinity

I have found that Figures 25.1 through 25.3 give students and researchers an intuitive feel for the batrachion, its self-similar structure, and Conway's contest. How might we apply other simple graphical tools to better display the behavior of the batrachions? I was particularly interested in how we can best illustrate the fractal (self-similar) structure of the various humps. One way is to renormalize each hump by shifting them on the x-axis each time a 2^k value is reached. In addition, the x-values can be scaled between 0 and 1 by multiplying by $1/2^k$. (This effectively normalizes the curves so the range 2^k to 2^{k+1} is squeezed into the range of 0 to 1.) Figure 25.4 illustrates my visualization and shows the beautiful but intricate relationships between successive humps. This plot was computed for the first 2,048 points in the batrachion, and I placed dots at the vertices of the first few humps. As the frog hops closer and closer to infinity, these curves become more squat and begin to hug the $y = 0.5$ line.

Interestingly, it is not clear how one hump in the batrachion is generated from the previous hump. As Colin Mallows of AT&T has pointed out, $a(100)$, which is located in the sixth hump, is computed as $a(a(99)) + a(100 - a(99)) = a(56) + a(44) = 31 + 26 = 57$. This shows that a point in hump 6 is generated from two points in

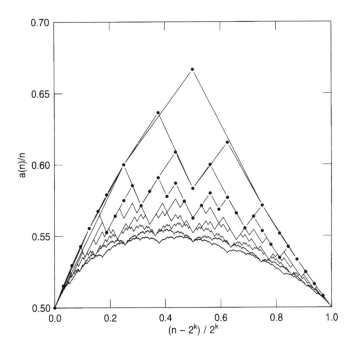

FIGURE 25.4. Renormalized batrachion to show fractal structure. (See text and program code in Appendix 1 for more information.)

hump 5 that are far apart. I have chosen one way to represent the complex relationship between the two terms by plotting $a(a(n - 1))$ (the first term of the batrachion formula) and $a(n - a(n - 1))$ (the second term) as x and y in an x-y plot (Figure 25.5). (The terms are normalized by dividing by n prior to plotting.) Notice the unusual clustering of points toward the bottom of the plot corresponding to the decay of the batrachion as it jumps to infinity.

Other Batrachions

When I first became interested in this class of curves, I thought there surely would be other famous batrachions. I was right. Figure 25.6 shows a batrachion computed from a very similar looking, recursive integer sequence:

$$Q(n) = Q(n - Q(n - 1)) + Q(n - Q(n - 2))$$

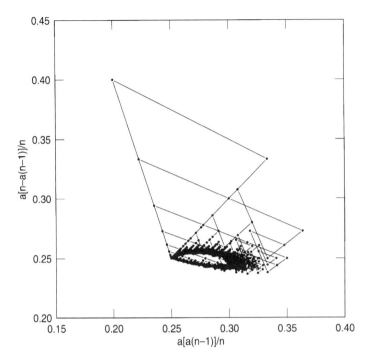

FIGURE 25.5. First and second terms of the batrachion as x and y.

This sequence was mentioned in Hofstadter's famous *Gödel, Escher, Bach*, but no figure was given to show its magnificent and bizarre behavior.[4] (The sequence of numbers produced by this formula is sometimes called Q-numbers.) The first few hundred values show striking regularity, but I do not know whether the regularities persist. There is obviously some subtle pattern that is difficult to define. The behavior of this function as it approaches infinity is not known. Does $a(n)/n$ eventually converge to 0.5? The virtual frog seems to be rather nervous, or perhaps high on LSD. There are periodic bursts that noisily decay to around 0.5 until another rapid hop starts again. How could such a simple integer formula give such complexity?

Finally, another batrachion, Mallows's batrachion, is defined by:

$$a(n) = a(a(n - 2)) + a(n - a(n - 2))$$

This also has an intricate humped structure. No one knows whether $a(n)/n$ tends to some limit for this batrachion.

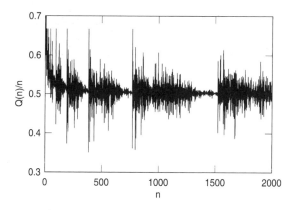

FIGURE 25.6. Visual representation of Hofstadter's batrachion.

The Crying of Fractal Batrachions

Various authors, such as Manfred Schroeder, have discussed how mathematical waveforms sound when converted to time waveforms and played as an audio signal. For example, Weierstrass curves (which are continuous but quite jagged) are a rich mine of paradoxes. They are produced by:

$$w(t) = \sum_{k=1}^{\infty} A^k \cos B^k t$$

where $AB > 1 + 3\pi/2$. If Weierstrass curves are recorded on a musical tape and replayed at twice the recording speed, the human ear will unexpectedly hear a sound with a lower pitch. Other fractal waveforms do not change pitch at all when the tape speed is changed. It is rumored (but I have not confirmed) that the first batrachion described in this chapter produces a windy, crying sound when converted to an audio waveform. I am interested in hearing from readers who have conducted such audio experiments on any of the batrachions. For other musical mappings of number sequences and genetic sequences to sound, see my book *Mazes for the Mind: Computers and the Unexpected.*[5] The BASIC and C programs in Appendix 1 are a launchpad for future exploration.

Recursive batrachians in amplexus.

CHAPTER 26

In a perfect universe, infinities turn back on themselves.
— George Zebrowski

Ramanujan, Infinity, and the Majesty of the Quattuordecillion

I have always been fascinated by the mathematical concept of nested roots. As background, most readers will probably be familiar with the commonplace square root symbol in mathematics:

$$\sqrt{4} = 2$$

Nested roots are certainly more fascinating (and typographically attractive) to look at. Here is an example:

$$\sqrt{\sqrt{4}} = 1.414 \ldots$$

If mathematics were a beauty contest, then nested roots would win the prize. They are simply square roots of roots of roots in an infinite cascade. Here is an example of an infinitely nested root:

$$\sqrt{2 + \sqrt{2 + \sqrt{2 + \ldots}}}$$

A few years ago I asked colleagues whether it were possible for humanity to determine the solution to the following nested set of square roots:

$$? = \sqrt{1 + Q\sqrt{1 + (Q + 1)\sqrt{1 + (Q + 2) \ldots}}} \qquad [1]$$

Here, Q indicates a quattuordecillion, or 1 followed by 45 zeros:

$$Q = 1,000,000,000,000,000,000,000,000,000,000,000,000,000,000,000$$

(Yes, *quattuordecillion* is the official name for this number in the United States. It comes from the Latin root *quattuordecim*. Check it out in your *Webster's Unabridged Dictionary*.) To get a feel for the size of a quattuordecillion, consider that it is much larger than the number of molecules in a pint of water (1.5×10^{25}) but less than the number of electrons, protons, and neutrons in the universe (10^{79}).

When I first became enamored by the number quattuordecillion I searched the world for other large number comparisons so I could adequately convey to students the majesty of the quattuordecillion. Here are some other comparisons. The Ice Age number (10^{30}) is the number of snow crystals necessary to form the Ice Age. The Coney Island number (10^{20}) is the number of grains of sands on the Coney Island beach. The talking number (10^{16}) is the number of words spoken by humans since the dawn of time. This last includes all baby talk, love songs, and congressional debates. This number is roughly the same as the number of words printed since the Gutenberg Bible appeared. The amount of money in circulation in Germany at the peak of inflation was 496,585,346,000,000,000,000. This number is approximately equal to the number of grains of sand on the Coney Island beach. The number of oxygen atoms contained in the average thimble is a good deal larger: 1,000,000,000,000,000,000,000,000,000. The number of electrons that pass through a filament of an ordinary light bulb in a minute equals the number of drops of water that flow over Niagara Falls in a century. The number of electrons in a single leaf is much larger than the number of pores of all the leaves of all the trees in the world. The number of atoms in this book is less than a googol. The chance that this book will jump from the table into your hand is not 0—in fact, using the laws of statistical mechanics, it will almost certainly happen sometime in less than a googolplex of years.

Numerical Sumo Wrestlers

I had read the mythical and mystic works, and it surprised me that I would be amazed when I finally reached down into the deepest part of my soul and found infinity. — Clark Punkford

When I asked colleagues to determine a value for the infinite set of nested roots in equation 1, I surely felt that no human could arrive at a solution. After all, a quattuordecillion is larger than most computers could assimilate. The behavior of the nested root also makes it particularly difficult to decipher. The continued roots seem to make the values smaller, while the Q multipliers continue to make the values larger. Think of two sumo wrestlers competing for domination on a numerical wrestling mat. How could we ever arrive at an amicable solution?

It turns out that it was not too difficult to compute the value for the mysterious nested root. The solution is so simple that it astonished most colleagues who expected a noninteger solution. The value of the nested Q system in equation 1 is simply $Q+1$, and this could have been solved decades ago without computers.

In 1911, a 23-year-old Indian clerk named Srinivasa Ramanujan posed the following question (#298) in a new mathematical journal called the *Journal of the Indian Mathematical Society*:

$$? = \sqrt{1 + 2\sqrt{1 + 3\sqrt{1 + \ldots}}}$$

Many months went by, and not a single reader could determine a solution. The difficulty was the infinite nesting of roots. Ramanujan finally threw up his hands and gave the answer, 3. It turns out that he had generated the problem years before in the form of a more general theorem:

$$x + 1 = \sqrt{1 + x\sqrt{1 + (x + 1)\sqrt{1 + \ldots}}}$$

Plug in any value you like for x, and the solution is $x+1$. Plug in Q, and the solution is $Q+1$. Try other values and impress your friends.

Absolute Reality

Ramanujan himself constructed a theory of reality using zero and infinity. (Most of Ramanujan's contemporaries in England could not understand what he was trying to say when he presented this arcane, mystical concept—can you?) To Ramanujan, zero represented "absolute reality." Infinity was the myriad manifestations of that reality. What happens if you multiply them together:

$$0 \times \infty$$

To Ramanujan, the product $0 \times \infty$ was not a single number, but all numbers, each of which corresponded to an "act of creation." What could Ramanujan have meant by this?

C H A P T E R **27**

---

*He watched her for a long time and she knew that he was watching her
and he knew that she knew he was watching her, and he knew that she
knew that he knew; in a kind of regression of images that you get when
two mirrors face each other and the images go on and on and on in some
kind of infinity.*

— Robert Pirsig, *Lila*, 1992

Recursive Worlds

Recursion

"Whatever can be done once can always be repeated," begins Louise B. Young in *The Mystery of Matter* when describing the shapes and structures of nature. From the branching of rivers and blood vessels to the highly convoluted surface of brains and bark, the physical world contains intricate patterns formed from simple shapes through the recursive application of dynamic procedures. Questions about the fundamental rules underlying the variety of nature have led to the search to identify, measure, and define these patterns in precise scientific terms.

Recursion is a fundamental concept in computer science, mathematics, biology, art (Figure 27.1), and even linguistics. For example, recursive definitions exist whenever something is defined in terms of itself. Consider the following scenario. You page through a dictionary to find the definition of recursion. Imagine you find it just below the word *recuperate*, as follows:

recumbent — lying down.
recuperate — to recover health.
recursion — look up the definition of recursion.

Here the word *recursion* is defined in terms of itself in a recursive definition. (Note that the term *recursion* is distinct from *iteration*, which is exemplified by turning the pages of the dictionary to find a particular word, which could be expressed in a computer program as: NextPage = CurrentPage + 1.) Here are some other linguistic recursive definitions:

A wolf pack is two wolves or is a wolf pack together with a wolf.

Even more simple constructs such as "Art is art," or "A dog is a dog," or "A dog is not a dog" might be considered primitive examples of recursive definitions. Sometimes the term *self-referential* is used when referring to these constructs.

The related concept of "quining" is discussed in Douglas Hofstadter's *Gödel, Escher, Bach*. Quining is a process of taking a group of words and forming a self-referential sentence by preceding the original group with the same group enclosed in quotation marks. Here is a concrete example:

"Is a sentence fragment of seven words" is a sentence fragment of seven words.

FIGURE 27.1.   Recursion in art. How many repeated levels do you see? ("Scruting the in-scrutable," from Roger Shepard's *Mind Sights.* © 1990 by R. Shepard. Reprinted with per-mission of W. H. Freeman and Co.)

Let me give a few interesting examples of recursion in different settings. Per-haps the most striking application of recursion occurs in the biological world, where growth starts with a bud, which grows into a pipe, which then branches into two buds, with each of these two buds branching in a recursive growth process. Iteration, or repeated application, of these simple rules results in a self-similar oak tree, or ar-terial blood system, or the bronchial system of lungs. The branching patterns are thought to be the result of the simplest of growth algorithms: The steps repeat the previous ones on smaller and smaller scales. Figure 27.2 shows a coral-like form I computed using simple recursive branching rules at smaller and smaller size scales (see program code in Appendix 1).

In geometry, one of the most interesting consequences of recursive processes is self-similarity. A self-similar object looks roughly the same after increasing or shrinking its size. Like a collection of nested Russian dolls, self-similar objects con-tain within themselves miniature copies of themselves. Look inside a turbulent stream of water: The largest eddies contain smaller ones, and these contain smaller ones still. The beautiful consequences of self-similarity are intricate, fine-grained patterns, now generally called fractals. The term *fractals* was coined in 1975 by Benoit Mandelbrot to encompass many of the detailed and convoluted shapes found in nature and produced by recursion in both the mathematical and natural worlds.[1]

One of my earliest and most interesting introductions to recursion and self-simi-larity came in the form of a cartoon by Don Martin in *Mad* magazine. In the first frame of the cartoon, a man lies anesthetized on a hospital operating table. Under the halogen lamps, the stainless-steel surfaces of the table glisten like diamonds. Blood moves through a clear, plastic IV line and into the patient's body through a vein in

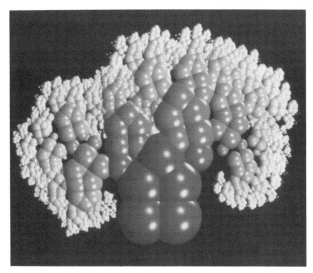

FIGURE 27.2.   Coral-like form generated from recursive program.

his thigh. One of the nurses bends close to the patient's face, checking his dilated pupils for signs of increased cranial pressure caused from excess fluid.

In the second frame of the cartoon, a brain surgeon uses a small circular saw to create an opening, a thin crack, all along the circumference of the man's head. The patient is pale and motionless. The doctor hands the medical saw to an assistant, opens the patient's cranial cavity by removing the skull cap, and then places this upper hemisphere of the skull in a bath of warm saline solution.

Inside the head, the surgeon does not find a brain, but rather another head, identical to the original and fully formed, but slightly smaller. The brain surgeon, slightly shocked, quickly removes this smaller head from the patient's skull cavity and begins to remove the skull cap of the smaller head. Inside the smaller head is yet another smaller head. He continues to find smaller and smaller heads until the last head, a few inches in length, sits in the palm of his hand. He opens this final head only to find a slip of paper about the size of a fortune in a Chinese fortune cookie. He holds the paper up. It reads, "Inspected by number 47."

An anthropological example of recursion occurs in the Ulupu tribe in New Guinea, where it is said that a musical ritual must be performed whenever a sad tune is played on a musical instrument. The musical ritual consists of three things: painting the face with certain botanical dyes, taking a deep breath, and playing a sad tune on a bamboo instrument. One can see why sad tunes are usually avoided.

A common visual analogy for recursion involves a video camera and a television (TV) monitor. Consider the following nomenclature: Camera(Cliff) means "Camera looking at Cliff Pickover," and TV(Cliff) means "TV shows a picture of Cliff Pickover." When operated in the usual manner, the camera produces an image of the subject as:

$Camera(Cliff) \rightarrow TV(Cliff)$

But if the camera is pointed at the TV, then, for a start,

$Camera(TV) \rightarrow TV(TV)$

However, the camera sees not just the TV, but the "TV showing a TV." Therefore,

$Camera(TV(TV)) \rightarrow TV(TV(TV))$

And now it sees "TV showing TV showing TV":

$Camera(TV(TV(TV))) \rightarrow TV(TV(TV(TV)))$

The feedback loop continues almost at the speed of light in a recursive process.

In computer programming, *recursion* is often used when referring to the structure and functioning of programs. The common definition is that a recursive program is one that calls itself, and a recursive function is one that is defined in terms of itself. Interestingly, recursion can be removed from any recursive program using iteration, as I shall show below.

Perhaps the most common example of recursion in programming and in mathematics (where recursion is called a recurrence relation) is the factorial function:

$$X! = X \times (X - 1)! \text{ for } X \geq 1, 0! = 1$$

where X is an integer.

The factorial funciton can be accomplished with a simple recursive program:

```
1 Factorial(X);
2    IF X=0 then Factorial=1
3    ELSE Factorial=X*Factorial(X-1)
4 END
```

Notice that the program calls itself in line 3.

Another common example of a recurrence relation is one that defines the Fibonacci numbers. This sequence of numbers, called the Fibonacci sequence after the wealthy Italian merchant Leonardo Fibonacci of Pisa, plays important roles in mathematics and nature. These numbers are such that, after the first two, every number in the sequence equals the sum of the two previous numbers:

$$F_N = F_{N-1} + F_{N-2} \text{ for } N \geq 2, F_0 = F_1 = 1$$

This defines the sequence: 1, 1, 2, 3, 5, 8, 13, 21, 34, 55 . . . . As with the factorial function, a simple recursive program can be written to generate the Fibonacci sequence:

```
1 Fibonacci(X)
2    IF N<=1 then Fibonacci=1
3    ELSE Fibonacci=Fibonacci(N-1)+Fibonacci(N-2)
4 END
```

In actuality, as with many recurrence relations, it is very easy to compute F_N using arrays in a nonrecursive program:

```
Fibonacci
    F[0]=1; F[1]=1;
    For i=2 to 30
            F[i]=F[i-1]+F[i-2]
    END
END
```

This program computes the first 30 Fibonacci numbers, using an array size of 30. This method of using arrays to store previous results is usually the preferred method for evaluating recurrence relations, because it allows even complex expressions to be processed in a uniform and efficient manner. Of course, in this example, the programmer can even avoid the array by retaining the last two values in two variables.

Recursive Lattices

Answer this question with "yes" or "no." Will your next word be "no"?
— Barry Evans, Everyday Wonders

The goal of the remainder of this chapter is to give examples of recursion for a particular class of self-similar objects I call recursive lattices, because they can easily be constructed using checkerboards of different sizes. The concept of repeatedly replacing copies of a pattern at different size scales for producing interesting patterns dates back to the 1890s, including the work of mathematicians Koch, Hilbert,[2] and Peano.[3] More recently work has been done by Mandelbrot and Lindenmeyer. Artists such as Escher, Vasereli, Shepard[4] and Kim have also experimented with recursive patterns. The designs in this chapter are so intriguing and simple to compute using a personal computer that I will give some computational recipes for readers who are computer programmers.

To create the intricate forms, start with a collection of squares called the initiator lattice or array. You can see what these look like in the magnification of the upper left corner in each lattice figure. The initial collection of squares represents one size scale. At each filled (black) square in the initial array I place a small copy of the filled array. This is the second size scale. At each point in this new array, I place another copy of the initial pattern. This is the third size scale. In practice I only use three size scales for computational speed and because a fourth size scale does not add much to the beauty of the final pattern.

In mathematical terms, begin with an S-by-S square array (A) containing all 0's to which 1's, representing filled squares or sites, are added at random locations. Here is an example:

```
0 0 0 0 0 0 0 0
0 0 0 0 0 0 0 0
0 0 0 1 1 1 0 0
0 0 0 1 0 0 0 0
0 0 0 1 0 0 0 0
0 1 1 1 0 0 0 0
0 0 0 0 0 0 0 0
0 0 0 0 0 0 0 0
0 0 0 0 0 0 0 0
```

Just how many patterns can you create by randomly selecting array locations and filling them with 1's? To answer this question, I like to think of the process of filling array locations in terms of cherries and wine glasses. Consider an S-by-S grid of beautiful, crystal wine glasses. Throw M cherries at the grid. A glass is considered occupied if it contains at least one cherry. With each throw a cherry goes into one of the glasses. How many different patterns of occupied glasses can you make? (A glass with more than one cherry is considered the same as a glass with one cherry in the pattern.)

It turns out that for an S-by-S array and M cherries, the number of different patterns is

$$\sum_{n=1}^{M} \frac{(S^2)!}{(S^2 - n)!n!}$$

As an example of how large the number of potential patterns is, consider that 32 cherries thrown at a 9-by-9 grid creates a more than 10^{22} different patterns. This is far greater than the number of stars in the Milky Way galaxy (10^{12}) and greater than the number of atoms in a person's breath (10×10^{21}). In fact, it is about equal to the estimated number of stars in the universe (10^{22}). For the patterns in this chapter, I use $S = 7$. Smaller arrays would lead to fewer potential patterns (particularly with the added induced symmetry, discussed later), and greater values of S lead to diffuse patterns with the scaling used.

In the C programming language, the process of filling the initial array can be coded as:

```
1  S=7; Sz=S-1; M=20;
2  for (h=1; h<=M; h++) {
3      /* rand returns a value between 0 and 32767 */
4      v = ((float)rand()/32767.)*S;
5      w = ((float)rand()/32767.)*S;
6      a[v][w]=1;
7      a[Sz-v][w]=1;
8      a[Sz-v][Sz-w]=1;
9      a[v][Sz-w]=1;
10 }
```

Notice that lines 6–9 introduce a four-fold symmetrical pattern that leads to an over-all symmetry in the design at several size scales. Because each of the "cherries" is symmetrically placed in each of the four quadrants of the initial array, each pattern is really defined by the smaller quadrant subpattern. This symmetrization decreases the number of possible patterns to a mere

$$\sum_{n=1}^{M/4} (((S+1)/2)^2)!/(((S+1)/2)^2 - n)!$$

For even values of S, use floor$((S+1)/2)$, where floor(x) returns the largest integer value not greater than the x parameter. I find that inducing the symmetry of the origi-nal pattern produces more esthetically appealing designs than those produced by a random initial array. Later in this chapter, I will mention other kinds of symmetry that can be explored.

For convenience, a computer program may store the values of the A array in two one-dimensional arrays, x and y, whose values are centered at the origin. The final value of N is the number of filled points in the symmetrical array:

```
N=0; Sz=S-1;
for (i=0; i<=S; i++)
   for (j=0; j<=S; j++)
         if (a[i][j]==1)
         {N++; x[h]=i-Sz/2; y[h]=j-Sz/2;}
```

The major computational step begins with the generation of the three size scales from these x and y arrays. The values of these arrays store the positions of the sites "filled" with 1s in the original structural motif. To determine the final positions of all the black squares in the final design, a scale factor α is applied to each position and the resulting terms summed:

$$X_i = \sum_{i=1}^{3} \alpha_i x_i$$

$$Y_i = \sum_{i=1}^{3} \alpha_i y_i$$

For the patterns in this chapter, $\alpha_1 = 1$, $\alpha_2 = S$, and $\alpha_3 = S^2$. In a computer program, this can be accomplished by:

```
for (i=1; i<=N; i++) {
   for (j=1; j<=N; j++) {
      for (k=1; k<=N; k++) {
            X = alpha[1]*x[h] + alpha[2]*x[h] + alpha[3]*x[h]
            Y = alpha[1]*y[h] + alpha[2]*y[h] + alpha[3]*y[h]
            PlotSquareAt(X,Y)
      }
   }
}
```

Beauty and the Bits

Several numerical measures of the patterns can be computed and displayed in an effort to quantify the structure of the patterns. If these parameters are found to correlate with the "beauty" of certain patterns, computer programs can then examine these parameters and automatically generate classes of patterns of aesthetic value. Imagine sitting back and letting your computer scan an "infinite" universe of forms and display only those you will find attractive.

Perhaps the most obvious parameter to compute is the fractal dimension D, which characterizes the size-scaling behavior of the pattern. This value gives an indication of the degree to which the pattern fills the plane, or how the pattern "behaves" through different magnifications. Luckily for computer programmers, the D-value for these recursive lattice patterns is easy to compute. In general, $N = S^D$ where N is the number of filled sites in the original symmetrical array of squares and S is the magnification factor used, in this case the same as the size of an edge of the original array, 7. Notice as the number of filled (black) sites, N, in the initial array increases, the dimension D increases. In one sense, D quantifies the degree to which the porous patterns fill the plane in which they reside. If all the sites in the initial array are filled, N is 49 and therefore D is 2. This makes intuitive sense: If the entire plane is filled, the dimension of the object should be 2. In a computer program, D is calculated from

```
D = log(N)/log(7)
```

In recursive words, dimensions are tangled up like a ball of twine, and all the patterns are neither one nor two dimensions but somewhere in between.

Another quantitative measure of the patterns' spatial characteristics is the length distribution function, $p(r)$. The function indicates the distribution of all the interpoint distances, r_{ij}, in the pattern. Mathematically speaking, this new function $p(r)$ is defined by

$$p(r) = (1/N)\sum\sum_{r_{ij}=r} 1$$

where N is the number of points in the pattern. The sums are over i and j, and what is being summed is $\delta(r_{ij},r) = 1$ if $r_{ij} = r$ and 0 if $r_{ij} \neq r$. In simple English, to compute $p(r)$, first select a point in the pattern and compute the distances from that point to every other point in the pattern. Do this for each point in the pattern, and create a graph showing the number of different lengths as a function of each length. When you interpret these graphs, the right-most point in the $p(r)$ graph is the maximum distance found in the structure, and the left-most point is the minimum. The most common distance is the highest point of the curve.

In a computer program, I use a Monte Carlo approach to compute $p(r)$, because actually computing all the interpoint distances is computationally expensive. Instead, one can randomly select pairs of points in the structure and produce the final $p(r)$ curve after 500,000 pairs are cataloged. In the C programming language, the Monte Carlo process looks like:

```
/* initialize p array to zero */
for (i=0; i<900; i++) p[i]=0;
total=500000;
/* catalog 500000 vector lengths */
for (i=0; i<total; i++) {
    rnd1 = ((float)rand()/32767.)*Npts;
    rnd2 = ((float)rand()/32767.)*Npts;
    xterm = corx[rnd1]-corx[rnd2];
    yterm = cory[rnd1]-cory[rnd2];
    dist = sqrt(xterm*xterm+yterm*yterm);
    dist = dist+.5;
    p[(int)dist]++;
}
```

R_{max}, the longest vector in the structure, or maximum linear dimension, can also be estimated by examining the last nonzero value of $p(r)$. The variable $Npts$ is the number of points in the final recursive lattice pattern. Obviously there is some noise in the Monte Carlo process, and I use a simple, nearest-neighbor smoother to reduce the noise so the eye can concentrate on the global structures of the curve:

```
for (i=1; i<900; i++) {
    ps[i] = .25*(float)p[i-1] + .5*(float)p[i] + .25*(float)p[i+1];
}
```

If you run the Monte Carlo process twice using different random numbers, the graphics look almost identical.

Another parameter, the radius of gyration, R_g, can be computed from the unsmoothed length distribution function $p(r)$ by

$$R_g = \sqrt{\frac{\int_0^\infty r^2 p(r)dr}{2\int_0^\infty p(r)dr}}$$

The radius of gyration quantifies the spatial extent of the structure in the plane. Small, compact patterns have small values of R_g. Large, extended patterns have large values of R_g. To get an intuitive feeling for this parameter, consider this: If all the squares in the final pattern were to lie on the edge of a circle of radius R, then $R_g = R$. If the circle were stretched in one direction to form an ellipse, the R_g value would increase because the "mass" of the pattern is further from the center of mass of the entire pattern. In the C language, the R_g computation looks like:

```
sum=0; summ=0;
for (i=0; i<900; i++) {
    summ=summ+p[i];
    sum=sum+p[i]*i*i;
}
Rg=sqrt(sum/((float)summ*2));
```

After examining hundreds of recursive lattice patterns, such as the full-page examples at the end of this chapter, I find that viewers often prefer high values of the fractal dimension of around 1.8, which is close to the average D for all the patterns in this chapter. The $p(r)$ curves for the preferred structures usually do not exhibit global features (bumps and valleys). To date, I have not been able to determine a correlation between perceived beauty and the R_g parameter. Readers may wish to compute the lattice patterns and also look for correlations between perceived beauty and the fractal dimension.

Higher Symmetry

As indicated earlier in this chapter, most people seem to prefer patterns with a symmetrical structure over those with purely random structures. I prefer patterns with four-fold symmetry; however, I have also experimented with patterns with inversion symmetry, bilateral symmetry, and random-walk symmetry. For example, at the end of this chapter you will see a few bilaterally symmetric patterns produced by making the left and right sides contain the same patterns:

$$A_{ij} = 1, A_{S-1-i,j} = 1$$

In C this is:

```
Sz=S-1;
for (h=1; h<=40; h++) {
    v = ((float)rand()/32767.)*S;
    w = ((float)rand()/32767.)*S;
    ar[v][w]=1;
    ar[Sz-v][w]=1;
}
```

Inversion symmetry can be computed by $A_{ij} = 1, A_{S-1-i,S-1-j} = 1$, or in a C program by:

```
Sz=S-1;
for (h=1; h<=30; h++) {
    v = ((float)rand()/32767.)*S;
    w = ((float)rand()/32767.)*S;
    ar[v][w]=1;
    ar[Sz-v][Sz-w]=1;
}
```

A few figures with inversion symmetry are also included. In addition, random walks can be used to force greater correlation between points in the initial array. Rather than selecting each site randomly, with random-walk symmetry the position of each new site is related to the previous site. For example, select a point and continually subtract 1 from or add 1 to the initial site's x- and y-coordinates. Other symmetries, such as the six-fold symmetry of a snowflake, can also be used.

Not all individuals to whom I showed the recursive patterns found them of esthetic interest. After seeing the 7-by-7 recursive lattice designs, Dr. Michael Frame of Yale University commented:

> As far as the aesthetics goes, I've been looking at fractal pictures for so long that these are too familiar to be very interesting. Berlyne elucidated a theory of perceptual arousal requiring two components: familiarity and novelty. When you've looked at self-similar patterns for a long time, the layers-upon-layers no longer hold any novelty. A less experienced audience might be of more use to you.

Digressions

Student: "What is the best possible question, and what's the best answer to it?"
Socrates: "You've just asked the best possible question, and I'm giving the best possible answer."

Readers are invited to experiment in several interesting areas:

1. Do the patterns you consider beautiful cluster around a single fractal dimension or radius of gyration value?
2. Are patterns with a bumpy $p(r)$ curve preferred to those with featureless curves?
3. Are patterns with larger starting arrays and greater size scales more aesthetically pleasing than those produced with the 7-by-7 arrays here?
4. Extrapolate the algorithm here to 3-D structures and higher-dimensional structures. How many different patterns can you produce in a 9-by-9-by-9 3-D initial array?
5. Perform a statistical test of correlation between aesthetic appeal and dimension or radius of gyration.
6. Generalize the recursive lattice program to nonsquare grids. For example, Figure 27.3 shows several patterns produced by Ken Shirriff of Oakland, California. He extrapolated the method to triangular grids.
7. Vincent Kruskal of New York sent me the following lyrics from the Shari Lewis show (*Lamb Chop's Play-Along*) as an example of recursion. He remarks, "One can see why sad tunes are usually avoided. Wisely she starts singing it at the *end* of her TV show."

This is the Song that Doesn't End
Yes, it goes on and on my friend
Some people started singing it
Not knowing what it was

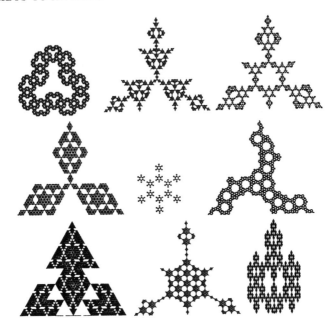

FIGURE 27.3.   Recursive lattice designs on a triangular lattice (see text).

And they'll continue singing it forever
Just because . . .
This is the Song, etc. etc.

Legal scholars generally agree that very limited excerpting from source material is not grounds for legal action. What does limited excerpting mean when applied to infinite recursive songs?

D = 1.712414
Rg = 143.7845

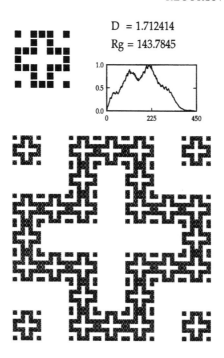

D = 1.654175
Rg = 144.0249

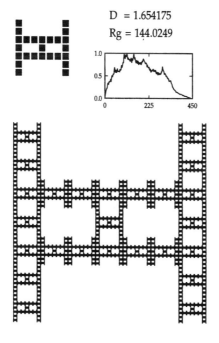

D = 1.812191

Rg = 145.1174

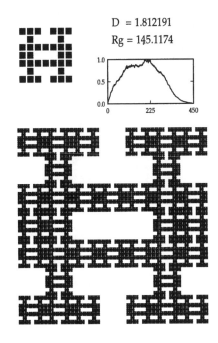

D = 1.693725

Rg = 146.9639

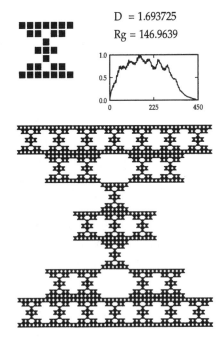

D = 1.796849
Rg = 124.4391

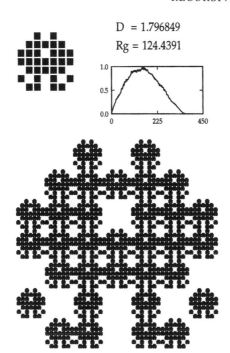

D = 1.841564
Rg = 140.6512

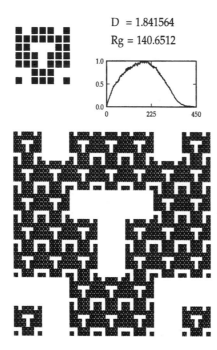

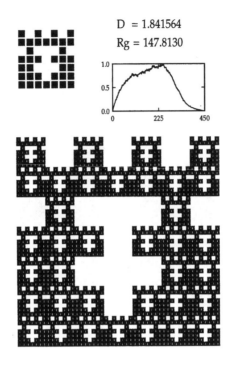

D = 1.841564
Rg = 147.8130

D = 1.882698
Rg = 146.2500

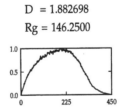

$D = 1.712414$

$Rg = 132.4023$

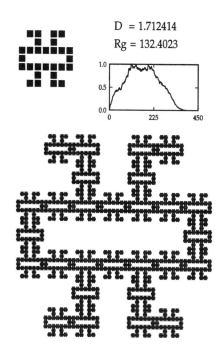

$D = 1.674330$

$Rg = 128.0688$

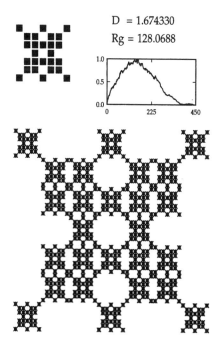

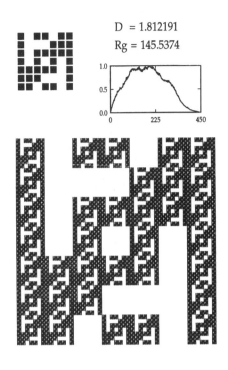

D = 1.812191

Rg = 145.5374

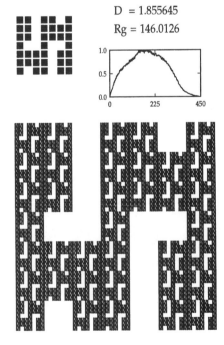

D = 1.855645

Rg = 146.0126

Macrozamia spiralis produced by methods similar to those discussed in chapters 6 and 23.

C H A P T E R 28

If the doors of perception were cleansed, every thing would appear to man
as it is, infinite. For man has closed himself up, till he sees all things
through narrow chinks of his cavern.

— William Blake, *Marriage of Heaven and Hell*

Chaos in Ontario

During a recent trip to the Ontario Science Center in Toronto, Canada, I came across an interesting puzzle. For those not familiar with the Science Center, it is located minutes from downtown Toronto. This vast playground of science has hundreds of exhibits inviting you to touch, try, test, and titillate your curiosity. The puzzle I saw there can be stated as follows. In the ten boxes below, write a ten-digit number. The digit in the first box indicates the total number of zeros in the entire number. The box marked 1 indicates the total number of 1's in the number. The box marked 2 indicates the total number of 2's in the number, and so on. For example, the 3 in the box labeled 0 would indicate that there must be exactly three 0's in the ten-digit number.

0	1	2	3	4	5	6	7	8	9	Row 1
3										Row 2

I watched a man and woman nearly have a fist fight as they argued about this problem while at the Ontario Science Center. The man said there was no possible solution. The woman screamed at him that he was wrong, that she had found a solution, and if he insisted on being so pig-headed about his ideas, she would leave him and take a cab from the Science Center. I left the couple behind, and never found out whether the woman left the man for more enjoyable company.

Who was right? Is there a solution to this problem? Are there many solutions to this problem? Before giving you the answers, I should first tell you that Daniel Shoham from MIT made some curious observations about the characteristics of row 2. He notes that because there are ten different digits in row 1, the sum of the digits in row 2 must be 10. He also derived a value for the maximum possible value for each digit of row 2. His logic is as follows. If digit x appears under digit y there must be x appearances of y, hence $x \times y < 10$. Therefore, the *maximum* number that can appear under each number is:

0	1	2	3	4	5	6	7	8	9	
9	9	4	3	2	1	1	1	1	1	Maximum

217

It turns out that that the woman in Ontario was right. There is a valid solution:

0	1	2	3	4	5	6	7	8	9	Row 1
6	2	1	0	0	0	1	0	0	0	Row 2

Notice that this number contains six 0's, two 1's, one 2, and so on. (Martin Gardner has written to me that 6210001000, which he calls the self-descriptive number, is the *only* solution.)

A more advanced problem, which I recently designed, continues this process to generate a sequence in a recursive fashion. In this variation, each row becomes a "starting point" for the next. For example, start with the usual 0 through 9 digits in row 1:

Row 1: 0 1 2 3 4 5 6 7 8 9

As I've just mentioned, 6210001000 is the solution to the puzzle:

Row 1:	0	1	2	3	4	5	6	7	8	9	R1
Row 2:	6	2	1	0	0	0	1	0	0	0	R2
Row 3:											R3

Now use row 2 as the starting point, and your next task is to form row 3, row 4, and so on, using the same rules. For example, a digit in the first space of row 3 would indicate how many 6's there are in row 3's ten-digit number. The second entry in row 3 tells how many 2's to expect in Row 2, and forth. Can you find a row 3 or row 4? Is it even possible to generate a row 3 or row 4?

Digressions and Contest

It turns out that row 3 and row 4 slip into an infinite cycle. In other words, using the values of 0123456789 for row 1, we get the following sequence: R1, R2, R3, R4, R5, R4, R5 . . . . As a contest, I asked colleagues to find the longest sequence of R's before they started repeating and entering a cycle. I also asked colleagues to start from an R1 of their choice!

The Portuguese Maneuver

Tomas Oliveira e Silva from Portugal found the solution to the first part of the puzzle (6210001000) using a C program running for 0.2 seconds on a DEC Alpha 3000 400 (see Appendix 2). A full solution is:

R1 = 0123456789
R2 = 6210001000
R3 = 0004440444
R4 = 6660006000
R5 = 0004440444

In an attempt to find the longest sequence of R's before they start cyclically repeating, Tomas examined all permutations of first-row entries. (This gives rise to solutions of the problem that are themselves permutations of the original solutions.) Therefore, to enumerate all essentially different cases he only had to generate one instance of each permutation of ten digits (it turned out that there are 92,378 different cases). An exhaustive solution was therefore within reach.

For each of these 92,378 cases, Tomas computed all solutions of the generalized problem and constructed a directed graph connecting the different cases. This project required about 4 hours on a DEC Alpha. Then, starting on each node, Tomas generated all possible paths ending in a loop. The longest of these paths *before* hitting a loop is:

R1 = 0000122245
R2 = 2222044430 (there are 3 other solutions)
R3 = 1111300023 (there is 1 other solution)
R4 = 3333411104 (there is 1 other solution)
R5 = 0000122241 (there is 1 other solution)
R6 = 2222044440 (there are 2 other solutions)
R7 = 0000800008
R8 = 2222022220
R9 = 0000800008 = R7 (loop begins again)

We currently do not know whether there are other paths of the same length. The longest length *loop* found is:

R1 = 0000111244
R2 = 1111444033 (there is 1 other solution)
R3 = 2222000311 (there is 1 other solution)
R4 = 0000111244 (there is 1 other solution)
R5 = 0000111244 (there is 1 other solution)

We currently do not know whether there are other paths of the same length. The longest path not leading to a loop, but rather to a dead-end, is:

R1 = 1223334444
R2 = 4223331111 (there are 42 other solutions)
R3 = 0332220000 (there are 4 other solutions)
R4 = 3550003333 (there are 2 other solutions)
R5 = 0555550000 (there is 1 other solution, leading to a loop)
R6 = No solutions (dead-end)

C H A P T E R 29

Mathematics is not a science—it is not capable of proving or disproving the existence of things. A mathematician's ultimate concern is that his or her inventions be logical, not realistic.

— Michael Guillen, *Bridges to Infinity*

To see the world in a grain of sand, and a heaven in a wild flower. Hold infinity in the palm of your hand, and eternity in an hour.

— William Blake, *Auguries of Innocence*

Cyclotron Puzzles

Cyclotron puzzles are fiendishly difficult, but, like many problems in mathematics and science, the rules of the game are really quite simple. In fact, you can study them using just a pencil and paper. These puzzles derive their name from the apparatus used for accelerating charged atomic particles as they revolve round and round in a circular orbit.

My interest in these kinds of puzzles that could be generated by computers, but solved using a pencil and paper, has continued since the publication of my book *Mazes for the Mind: Computers and the Unexpected.* Cyclotron puzzles are entirely new, and played on an annular board filled with random numbers from 0 to 100. Figures 29.1, 29.2, and 29.3 show typical examples. Each "site" on the board contains a single-digit number or a two-digit number. The code in Appendix 1 shows how you can generate an infinite number of boards using a C program.

A cyclotron particle may start on any number on the board, and your job is to find longest possible path through the board by moving horizontally or vertically (not diagonally) through adjacent squares. This means the particle takes a single step (up, down, right, or left) during each movement. There are two additional constraints as follows:

```
 2   3  11  84  10  92  63  72  19  91  98  68  51  16  46  77  14  12  46  63
23  51  26  34  73  94  27  49  73  98  60  44  36  31  79  73  67  72  56  74
11  71  40  25  22  31  83  31  20  96  23  96  74   3   6  13  97  87  25  33
87  92  73                                                      79  50   3
45  57  61                                                      33  55  81
23  48  43                                                      85  50  28
73  42  29                                                      39  97  92
56  31  61                 Cyclotron 1                          17  23  19
88  40  52                                                      13  32  71
54  79  11                                                      51  56  49
 9  60  43                                                      11  99  47
99  13  20                                                      34  12  32
12  48  26  67  37  34  49  56  99  32  39  94  11  23   9  29  45  56  62  65
90  70  70  15  25   6  44  77   8  66  14  54  93   3  78  95  99  99  18  69
13  20  62  53  61   6  82  55  43  79  98  37  46  26  97  66  43  49  25  64
```

FIGURE 29.1.

```
26 32 75 62 99 82 10 83 27  9 62  8 69 62 15 31  1 47 73 67
80 61  7 47 25  9 27 53 18 70 83 64 22 51 31  9 15 23 74 93
29 54 13 11 11 47 16 92 64 94 42  9 17 51 22 80 71 50 41  7
55 51 27                                          60 45 22
99 40 72                                          96 16  9
59 65 99                                          54 82 38
89 29 84                                          63 56 18
 3 91 49              Cyclotron 2                 92 34  9
97 18 17                                          51 20 21
87 63 61                                          60 19 87
97 50 49                                          48 26 17
69 24 55                                          77 65  7
52 48 37 19 71 11 29 94 79 38 91 79 85 25 65 63 12 33 80 37
27 23 42 34 97 32 64 85 75 60 35 45 24 11 71 82 61 78 48 16
25 82 60 88 69 86  9 45 62 85 60 28 14 73 66 42  0 25 12 98
```

FIGURE 29.2.

1. Each number along the particle's path must be different; that is, you can use each number only once along a particle's path.
2. Your particle may only travel in an all-clockwise or all-counterclockwise direction. In other words, the particle must go round and round in one direction, but it can orthogonally switch between the three tracks as useful.

Heiner Marxen from Berlin has found what he thinks are the longest path solutions, and I have included these in the "Solutions" section in this chapter. No one knows the shortest path length for any of these puzzles. (By path length, I mean the number of sites through which a path travels.) I am interested in any attempts by readers at finding long or short paths. Feel free to try to find such paths using only a pencil and paper. Challenge a group of friends to search for the longest path they can find within 15 minutes!

```
48 29 44 18 90 91 69 25 62 98 11  4 39 33 64 81 46 95 62 38
68 24 96 18 94 87 14 17 32 69 89 69 27 25 88 67 11 21 32  6
44 42 69 11 35 74 59 43 29 45 99 13 53 42 63 30 81 80 13  7
38 31 43              1                            62 42 28
77 67 86   Cyclotron 3   2     "Leaky             90 33 78
 2 38  0              3     Cyclotron"            57  9 90
 7 88 72              4                            20 81 34
92 24 47 10 11 12 13 14 15  5 16 17 18 19 20 21 22 38 82  5
90 11 53              6                            89 85 94
13 45 75              7                            49 22  0
61 25 34              8                            68 35 93
42 35 68              9                            24 64 89
54 37 46 83 30 33  4 83 87 18 96 53 45 18 50 64 61 98 18 44
68 89 35 74 36 60 36 95 41 31 92  4 22 80 16 11 20 43 19 28
66  3 46 50 29 52 81 32 76 45 57 89 56 20 47 34 78 94 82 88
```

FIGURE 29.3

I have included three different boards in this book. Board 3, which I have called a "leaky cyclotron," has a slight modification, because you can use the central tracks as needed when finding either the longest or shortest path. (In other words, while in the central channels, you can travel in either direction, but the constraint of having each number along a particle's path be unique persists.)

Some colleagues have noted that the definition of clockwise or counterclockwise movement is difficult to understand when a particle is travelling through the corners of the board. As a clarification, the following diagram shows permissible moves from a starting position, marked with a capital letter, to various neighbor positions, marked by a corresponding lowercase letter. These moves would all be considered valid if you were to restrict the particle to clockwise motion.

```
-A -a -- -- -- -c -- -- -- -- -- -- -- -- -- -- -- -- -d --
-- -b -- -- -- -C -c -- -- -- -- -- -- -- -- -- -- -- -D -d
-b -B -b -- -- -c -- -- -- -- -- -- -- -- -- -- -- -- -d --
-- -- --                                              -- -- --
-- -- --                                              -- -- --
-- -- --                                              -- -- --
-- -- --                    Some valid moves          -- -- --
-- -- --                    for "clockwise"           -- -- --
-- -- --                    motion                    -- -- --
-- -- --                                              -- -- --
-- -- --                                              -- -- --
-- -- -e                                              -- -- --
-- -e -E -- -- -- -f -F -- -- -- -- -- -- -- -- -- -- -- --
-- -- -e -- -- -- -- -f -- -- -- -g -- -- -- -- -- -- -- --
-- -- -- -- -- -- -- -- -- -- -g -G -- -- -- -- -- -- -- --
```

Heiner Marxen notes that this permits cycles of the following kind in the corners:

```
5  6  7 -- -- -- -- -- -- -- -- -- -- -- -- -- -- -- -- --
4  -- 8 -- -- -- -- -- -- -- -- -- -- -- -- -- -- -- -- --
3  2  1 -- -- -- -- -- -- -- -- -- -- -- -- -- -- -- -- --
-- -- --                                       -- -- --
-- -- --                                       -- -- --
-- -- --                                       -- -- --
-- -- --                                       -- -- --
-- -- --                                       -- -- --
-- -- --                                       -- -- --
-- -- --                                       -- -- --
-- -- --                                       -- -- --
-- -- --                                       -- -- --
-- -- -- -- -- -- -- -- -- -- -- -- -- -- -- -- -- -- --
-- -- -- -- -- -- -- -- -- -- -- -- -- -- -- -- -- -- --
-- -- -- -- -- -- -- -- -- -- -- -- -- -- -- -- -- -- --
```

Of course, the rules that do not permit reuse of grids would make these cycles of very limited use when searching for long paths. They may have some value at the start or finish of a path.

Cyclotrons to Infinity

Consider the following questions involving the cyclotron puzzle:

1. Can you solve this class of problems using pencil and paper?
2. Can you solve this class of problems using a computer program?
3. How many different unique particle tracks would you expect to find in cyclotron boards of the size in this chapter?
4. Extend the puzzle to 3-D worlds where particles tunnel through doughnut-like structures. Use computer graphics to display the longest paths as the computer finds them.
5. Explore huge cyclotron worlds containing thousands of locations.
6. How would the kinds of solutions (and difficulty of finding solutions) change as the board sizes approach infinity?
7. Given a set of cyclotron worlds constructed randomly as in this chapter, what is the *average* largest cyclotron path you would expect to find?
8. Is it better to start your path at a particular place in the board? In other words, do certain regions give rise to longer paths than others?
9. In true cyclotrons, the orbit of a particle cannot decrease in diameter. For the puzzles here, this additional constraint would imply that a particle cannot jump to a channel of smaller diameter as it travels. It could only travel in its current channel or go to an outer one. How would this alter your answers to the other questions?

Solutions

Since I originally designed and presented these puzzles, Heiner Marxen from Berlin has become the world's foremost cyclotron puzzle solver. (See Appendix 2 for Heiner's program code.) Although you do not need a computer to solve these problems, Heiner used a program that required around three hours to solve each cyclotron running on an HP9000/710 workstation. His solutions are schematically illustrated in this section. We believe these are the best solutions for each cyclotron. Can you find longer solutions? To arrive at these "longest" solutions Heiner's program performed what we believe is an exhaustive search, but we do not know whether there are other solutions of the same length. We also do not know the shortest path for each puzzle.

The maximal path length for cyclotron 1 is 70. In the schematic illustration of the path, the first position of the sequence is marked 1; the second 2; and so on. The last position is marked 70. Assuming the upper-left corner to be $(1, 1)$ and the lower right to be $(20, 15)$, then this sequence starts at $(6, 14)$ and ends at $(20, 15)$. The first few numbers on the path are 6 - 34 - 37 - 25 - 15 - 70 - 26 - 20 - 43 - 60 - 9 - 54 - .... Because 54 is the 12th number in this sequence, its position, $(1, 10)$, is marked 12 in the solution diagram.

```
-- 26 27 28 29 -- 33 34 35 36 37 38 -- 44 45 46 47 -- -- --
-- 25 -- -- 30 31 32 -- -- -- -- 39 42 43 -- -- 48 -- -- --
-- 24 -- -- -- -- -- -- -- -- 40 41 -- -- -- 49 50 -- --
-- 23 --                                     51 -- --
21 22 --                                     52 53 54
20 19 --                                     57 56 55
-- 18 17              Cyclotron 1            58 -- --
-- -- 16             solution               59 -- --
13 14 15                                     60 61 –
12 -- --                                     -- 62 63
11 10  9                                     -- 65 64
-- --  8                                     -- 66 --
-- --  7 --  3  2 -- -- -- -- -- -- -- -- -- -- -- -- 67 68
-- --  6  5  4  1 -- -- -- -- -- -- -- -- -- -- -- -- 69
-- -- -- -- -- -- -- -- -- -- -- -- -- -- -- -- -- -- 70
```

Cyclotron 2: Path length 73

```
-- -- -- -- -- -- -- -- -- -- --  1  2  3  4  5  6  7 -- --
-- -- -- -- -- -- -- -- -- -- -- -- -- -- -- -- --  8  9 10
-- 73 72 -- -- -- -- -- -- -- -- -- -- -- -- -- -- -- -- 11
-- -- 71                                      -- -- 12
68 69 70                                      -- -- 13
67 -- --                                      -- 15 14
66 65 64              Cyclotron 2             -- 16 --
61 62 63             solution                 18 17 –
60 59 --                                      19 20 21
-- 58 --                                      -- -- 22
-- 57 --                                      -- 24 23
-- 56 55                                      26 25 --
-- -- 54 53 52 51 -- 47 46 -- -- -- -- -- -- -- 27 28 --
-- -- -- -- -- 50 49 48 45 44 43 42 -- -- -- -- 35 34 29 30
-- -- -- -- -- -- -- -- -- -- 41 40 39 38 37 36 33 32 31
```

Cyclotron 3: Path length 71

```
-- -- -- -- -- -- -- -- -- -- -- -- -- -- -- -- -- --  1 --
-- -- 52 53 54 55 -- -- -- -- -- -- -- -- -- -- -- --  2  3
49 50 51 -- -- 56 57 58 59 60 -- -- -- -- -- -- -- --  4
48 -- --              61                      -- --  5
47 46 45              62 "Leaky cyclotron"    -- --  6
-- -- 44              63 solution             -- 8  7
-- -- 43              64                      -- 9 --
-- -- 42 71 70 69 68 67 66 65 -- -- -- -- -- -- -- -- 10 --
-- -- 41              --                      -- 11 --
-- -- 40              --                      13 12 --
-- 38 39              --                      14 -- --
-- 37 --              --                      15 -- --
-- 36 35 34 33 32 -- -- -- -- -- -- -- 19 18 17 16 -- --
-- -- -- -- -- 31 30 29 28 27 26 -- -- 21 20 -- -- -- -- --
-- -- -- -- -- -- -- -- -- -- 25 24 23 22 -- -- -- -- -- --
```

Vampire Numbers

Anne Rice

If we are to believe best-selling novelist Anne Rice, vampires resemble humans in many respects, but live secret lives hidden among us mortals. Consider a numerical metaphor for vampires. I call numbers such as 2,187 "vampire numbers" because they are formed when two progenitor numbers, 27 and 81, are multiplied together ($27 \times 81 = 2,187$). Note that the vampire, 2,187, contains the same digits as both parents, except that these digits are subtly hidden, scrambled in some fashion. Similarly, 1,435 is a vampire number because it contains the digits of its progenitors, 35 and 41, ($35 \times 41 = 1,435$). These vampire numbers secretly inhabit our number system, but most have been undetected so far. When I first considered vampire numbers, I believed there were only six, four-digit vampires in existence and had no idea whether there were any larger vampire numbers.

True Vampires versus Pseudovampires

"True" vampires, such as in the two previous examples, have an even number of digits, and each one of the multiplicands has half the number of digits of the vampire. Of most interest to me are true vampires that are not created by simply adding 0's to the ends of numbers, as in: $270,000 \times 810,000 = 218,700,000,000$.

On April 29, 1994, I challenged computer scientists and mathematicians around the world to submit the largest vampire number they could find. It turns out that there are certain situations in which arbitrarily large, "pseudovampire" numbers can be found. Pseudovampires are numbers that have the same properties as vampires, but the two multiplicands have a different number of digits. After reading my chal-

227

lenge, Dean Hickerson from California found that one could place any number of zeros in certain numbers to construct huge pseudovampire numbers. Below is one example:

$$2 \times 600000031578947368421 = 1200000063157894736842$$

In other instances one can place an arbitrary number of 9's:

$$3 \times 499999997931 = 1499999993793$$

Amazingly, one can also continually increase the lengths of the contiguous runs of 3's and 6's by the same amount or increase the lengths of the blocks of 1's, 2's, 4's, 5's, 7's, and 8's by the same amount in:

$$14 \times 90088877776666555544433322221111$$
$$= 1261244288873331777762066511095554$$

Dean used the software Mathematica to perform the multiplications and count the digits.

James Allen notes that if $x \times y = z$ is a pseudovampire, then so is $x0 * y0 = z00$. Using a C program running on a Sparc machine, he also found the three largest pseudovampires less than 2^{32}. They are $4496 \times 955286 = 4294965856$, $9489 \times 452625 = 4294958625$, and $43365 \times 99042 = 4294956330$.

Upon hearing of my interest in vampire numbers, Tony Davie from Scotland found other four-digit pseudovampires. He wrote the following Gofer program to perform the search:

```
vamps = [(x,y,z) | x <- [1..[, sx=show x, y <- [1..x[,
            z = x*y, sort(show z)== sort(sx++show y)[
```

(See Appendix 2 of this book for an interpretation of this little Gofer program and for more information on Gofer.) Using this approach, Tony found other four-digit pseudovampires: 1,435, 1,530, 1,260, 2,187, 6,880, 1,827, 1,395, 1,206, 1,255. Of these, the following are true four-digit vamps: $21 \times 60 = 1,260$, $15 \times 93 = 1,395$, $35 \times 41 = 1,435$, $30 \times 51 = 1,530$, $21 \times 87 = 1,827$, $27 \times 81 = 2,187$, and $80 \times 86 = 6,880$.

Dick Adams from Maryland suggests that we refer to the largest vampire number as "vampire omega."

Steve Frye found 155 six-digit true vampires using the programming language REXX (see Appendix 2 at end of book). It required about 20 minutes using a 486DX 33MHz PS/2 running OS/2 2.1. The largest six-digit vampire is: $896 \times 926 = 829,696$.

Finally, Dean Hickerson notes that if we allow researchers to add zeros to the interior of numbers, large vampires can be created, as in:

$$1100000001110879 \times 9100000000000001$$
$$= 10010000010109000000000001110879$$

In general, Dean uses Mathematica to search for vampire numbers. Below is a chunk of a Mathematica program:

```
dc[n_]:=Block[{ct={0,0,0,0,0,0,0,0,0,0},k=n},
  While[k>0,
    ct[[Mod[k,10]+1]]++;
    k=Floor[k/10] ];
  ct]

m[a_,b_]:=Block[{},
  Print[dc[a b]-dc[a]-dc[b]];
  Print[a b];
  If[dc[a]+dc[b]==dc[a b],
    Print["**********************************"]]
]
```

In this code, $dc[n]$ returns a vector showing how many of each digit are in n. The variable $m[a,b]$ prints the excesses (which may be negative) of each digit in the product ab with respect to those in a and b, the product itself, and a line of asterisks if the product is vampiric.

Cliff Reiter from Lafayette College's Mathematics Department is one of the world's foremost vampire number experts, and he has made some additional discoveries regarding the vampires. To perform the search, he used the programming language J, which uses ASCII symbols, sometimes with a dot or colon as a suffix, to designate common mathematical functions.[1] Similar to what I discussed previously, he also found that vampire numbers can be extended by adding a trailing 0. For example, starting with $27 \times 81 = 2,187$ we can create $27 \times 810 = 21,870$, thus it would seem there are vampire numbers of all sizes. He also found other ways of adding zeros to the interior of numbers, as in $251 \times 4,001 = 1,004,251$, $251 \times 40,001 = 10,040,251$, and $251 \times 4,000,001 = 1,004,000,251$. Cliff was the first to find the true vampires $231 \times 534 = 123,354$; $231 \times 543 = 125,433$; $231 \times 588 = 135,828$; $231 \times 750 = 173,250$; and $231 \times 759 = 175,329$.

Most important, Cliff has the distinction as having found the largest true vampire number to date, excluding those that can be formed by adding more and more zeros to the interior of numbers:

$$1234554321 \times 9162361086 = 11311432469283552606$$

He also found another, nearly identical, 20-digit vampire:

$$1234554321 \times 9162291363 = 11311346392452629523$$

Notice that this number has a shared vampire factor with the previous and does not contain any zeros. Can you beat these 20-digit monstrosities? They required a few hours of computation on a 486 PS/2.

Cliff also found other very large vampires:

$1234569 \times 9116655 = 11255139646695$
$12345678 \times 94919814 = 1171849459463892$
$123456789 \times 916494741 = 113147497859246649$
$12381 \times 90768 = 1123798608$
$123452 \times 914837 = 112938457324$

$123453 \times 912051 = 112595432103$
$1234569 \times 9116277 = 11254672979613$

Steve Frye found other 12-digit vampires:

$100151 \times 998609 = 100011689959$
$100211 \times 999872 = 100198172992$
$100211 \times 998063 = 100016891293$

Cliff has also found some small triple pseudovampires, such as $28 \times 71 \times 9 = 17,892$. The largest triplet, with no zeros, is $29 \times 75 \times 91 = 197,925$. Interestingly, Dean Hickerson has found a six-digit vampire ($125,460 = 204 \times 615 = 246 \times 510$) and 18 eight-digit vampires that have two pairs of factors each. Three eight-digit vampires have one vampire as a factor:

$10081260 = 8001 \times 1260 = 8001 \times 21 \times 60$
$17429580 = 9540 \times 1827 = 9540 \times 21 \times 87$
$17988642 = 9846 \times 1827 = 9846 \times 21 \times 87$

None are the product of two vampires.

For those readers curious about the "look and feel" of the programming language Cliff used, here is an example session from J.

```
dig=.#&10@>.@(10&¬.)@>: #: [   NB. gives the digits as vector
dig 2345
2 3 4 5
vampq=./:~@dig@* -: /:~@,&dig NB. Vampire query
27 vampq 81                   NB. Vampire pair? Yes
1
26 vampq 81                   NB. Vampire pair? No
0
NB. vampa looks for vampire associates of numbers on left
NB. from the list on right
vampa=.(*@{.@}.@|:#[)@:((L,vampq"0 #[)"0 1)
27 vampa i.100                NB. only associate of 27 among 0-99
27 81
27 vampa i.1000               NB. only associates of 27 among 0-999
27 81 810

(i.100) vampa i.100
```

Vampires and the Infinite

One must say then that the mass of seeming irrationalities, anomalies, contradictions, and proliferating infinities may be the most important parts of any science. They drive the accumulation of knowledge.
— George Zebrowski

As the numbers get larger and larger, how often do you expect to find vampires? Do they become more secretive (sparser) or more outgoing (frequent) as one searches for vampires up to a googol?

James Allen has studied such questions for pseudovampires. In fact, using a C program running on a Sparc machine, he has found 1,064 pseudovampires among the 12.5 million integers $x \times y$ where ($5,000 \leq x \leq y \leq 9,999$). This corresponds to 17 pseudovampires per 200,000 numbers. Next he generated 200,000 random 4-tuples (x,y,x',y') for the same range and checked for vampireness using the "test" product $x \times y \equiv x' \times 10,000 + y'$. He found 17 of these "test" vampires. Thus he conjectures that vampireness occurs at roughly the same rate one would expect if the digits of a product behaved like uniform random variables.

Glenn S. Knickerbocker from New York discovered a property of vampires that speeds computer searches for these numbers. It turns out that the sum and product of the digits are equal, modulo 9. Because this is only true for (0, 0), (2, 2), (3, 6), and (5, 8), only 6 out of every 81 pairs are possible vampires. His search program, running on an IBM VM operating system, found the four-digit and six-digit vampires in less than a minute. Using his REXX program (see Appendix 2 at end of book), he finds that there are:

```
   7   four-digit vampires
 155   six-digit vampires
3382   eight-digit vampires
```

Even though there are more vampires in each log range, vampires are actually becoming less dense in each range. (Is this strange considering that large numbers have more factors than smaller ones?)

Similarly, Stephen H. Landrum from California suggests that the ratio V_n/n (where V_n is the number of vampire numbers up to n) will in general decrease as n gets larger, but the ratio $V_n/log(n)$ will increase as n gets larger.

Dean Hickerson suggests a simple heuristic argument for determining how secretive the vampires become through the number system. If we multiply two random numbers with a total of n digits, their product will have the same number of 0-digits, 1-digits . . . , and 9-digits as the factors with probability $C/n^{9/2}$ for some constant C. This suggests they should become more secretive (sparser) as one climbs the ladder of the integers toward a googol. With enough work, one could probably determine the exact value of C, but Dean thinks that proving that vampires are increasingly secretive would be nearly impossible.

Digressions

What is the largest true vampire number you can find lurking out there in the world of integers? Glenn S. Knickerbocker from New York notes that no vampires or pseudovampires are products of two vampires or pseudovampires with six or fewer digits.

Dean Hickerson asks: What's the smallest vampiric product of the form $N \times N = N^2$? In other words, which N^2 has twice as many of each digit as N? (Heuristically, there should be infinitely many such N.)

C H A P T E R 31

Sequences of truly patternless, unpredictable digits are a valuable commodity.

— Ivars Peterson

Anyone who considers arithmetical methods of producing random digits is, of course, in a state of sin.

— John von Neumann

Computers, Randomness, Mind, and Infinity

Humans Cannot Produce Random Numbers

Sit down at a computer keyboard, or a typewriter, and randomly press the 1 and 0 keys. Try to make the string of numbers as "patternless" as possible. In other words, try to generate random numbers. Here is my first attempt, typing 1 with my left hand and 0 with my right hand:

```
1011100011100010101101001010010100101000101001110101100011010100
0111010101011101110101011010101010001010111011100101010101000101000
1010111010101110111010101001010101010010101011101001001000100010010
0010010101000100101011000101011001010011101000100011110101010101010
1010111010101111011010101011100101010100100010101001010101010101010
1011101010010010000101010101110101011
```

Before reading further, try this experiment on yourself and friends. I have tested dozens of supposedly random sequences typed by colleagues, and found out that it is curiously difficult for humans to type patternless sequences. To start with, we would expect about a 50 percent occurrence of each digit, and I did amazingly well. The above example had 49.3 percent 1's and 50.7 percent 0's. Next, we would expect 25 percent occurrences of the following pairs: 00, 11, 01, 10. In fact, doing my best to make a random sequence, I produced 15 percent, 13 percent, 36 percent, and 36 percent occurrences, respectively. Apparently my fingers preferred to oscillate rather than producing doublets such as 00 and 11. Perhaps I was trying to avoid clumpiness of digits, when in fact strings of identical digits should exist in a truly random sequence. I strongly suspect that if I were to analyze all possible triplets of digits, my fingers (and medulla oblongata) also did not behave very randomly.

Do left-handed and right-handed people produce different kinds of sequences? As one tires of typing, does the randomness increase or decrease? Could the statistics for different individuals be used to diagnose any psychological or neurological diseases? Do the percentages tend to characterize, or "fingerprint," a particular individual? These are all unanswered questions.

On a similar topic, Manfred Schroeder (*Fractals, Chaos, Power Laws*, 1991) discusses a machine called "Shannon's outguessing machine" that analyzes the random head-tails choices of a human contender. In this gambling game, a player selects a head or tail, and a computer attempts to guess the player's selection. If the computer guesses correctly, the human loses money; otherwise, the human wins money. Does a player change after losing a throw? Does the player keep on choosing tails if tails has brought two previous wins? Or does the gambler get nervous and select heads next? For humans, such strategies are mostly subconscious, but the machine can uncover the hidden correlations in people's selections and will beat any human after playing many games. Humans simply are not capable of producing random heads or tails, 0's or 1's. Our minds abhor long strings of like outcomes—as would occur naturally in truly random sequences.

The Lure of Random Numbers

In modern science and computer art, random number generators have proven invaluable. The generation of random numbers has a long history. Since antiquity, societies have had a need to generate and use "chance" events in a variety of settings. Archeologists know that prehistoric humans used dice much like ours, played dice games similar to our own, and cheated opponents by using loaded dice! Before dice became useful in games, cubes with numbers were used as magical aids for divining the future. (Astragalomancy is the practice of divination by means of dice.) Primitive humans probably used cubical knucklebones or the anklebones of sheep as dice-like game pieces.

In modern science, random number generators are useful in simulating natural phenomena and in sampling data (see various sources in "For Further Reading"). They are used by scientists for tackling a wide variety of problems, such as developing secret codes, modelling the movement of atoms, and conducting accurate surveys.

How difficult is it to build a random number generator using a super-simple computer program? How can we diagnose the quality of the random number generator with simple graphical tools? My goal in this chapter is to stimulate and encourage programmers, students, and teachers to explore a few simple techniques. You can also consider the colorful representations solely as art objects, providing an example of the link between mathematics and art (as do other chapters in this book).

Randomness

In a sense, randomness, like beauty, is in the eye of the beholder.
— S. Park and K. Miller, 1988

The prevailing view of researchers working with computers is that it is very difficult to casually create a good random number generator (RNG) using a simple algorithm in a computer program. In fact, if you were to tell colleagues that the first function that came to your head would make a fine RNG for most applications, surely they would shake their heads with disbelief at such heresy. Many papers and books passionately warn of the complexities of creating random numbers. (For an excellent review, see Park's and Miller's article "Random Number Generators: Good Ones are Hard to Find," in "For Further Reading.") I have found that wise caution has sometimes led to an almost mystical, exaggerated sense of difficulty in this area.

As I alluded to in the previous section, many years of experience have demonstrated the usefulness of RNGs in computer programming, numerical analysis, sampling, simulation, recreation, and decision making. There has also been considerable interest in the area of computer graphics, where random numbers are used to construct complex and natural textures, terrains, and various special effects.

I begin our discussion by defining what I mean by a "good" random number generator. (Rather than simply producing 1's and 0's as in our previous typing test, these generators produce a range of numbers between 0 and 1.) First, I require that the RNG produces a few million numbers that are statistically independent and uniformly distributed between 0 and 1. Second, I have the interesting requirement that the generating formula and parameters be easy to remember. By far the most popular RNGs today make use of the linear congruential method and generally have the form:

$$X_{n+1} = (aX_n + c) \bmod m, \quad n \ge 0.$$

Here, a is the multiplier, m the modulus, c the increment, and X_0 the starting value. (In modular arithmetic, if the number is greater than $n - 1$ it is divided by n and the remainder used in place of the ordinary result.) Usually, specific, multidigit numbers are required for these variables, and I have not found (even among the most seasoned computer scientists) a single person who actually could write down the parameters for one of these RNGs without consulting textbooks. Many pages have been written on the theoretical reasons for using this number or that for the modulus, multiplier, and so on. Fascinating folklores have even evolved regarding some initial seeds that are better than others. I recall, along with many colleagues, being told to use a seed with at least five digits, with the last digit being odd. This folklore is confirmed by Park and Miller, who note, "For some generators this type of witchcraft is necessary."

Because many computer scientists and students will never have more than an occasional need to use a random number generator, and on most occasions superlong periods and statistical perfection are not of paramount importance, I would like to begin with an anecdote regarding an easily remembered RNG with interesting properties. There are many areas in which "good" RNGs (by my definition) are quite useful, and I am also interested in their use in computer graphics.

The Cliff RNG

On February 1, 1994, I posted the following RNG to various computer bulletin boards. I called it the "Cliff" RNG, and wanted to impress researchers with how easy it is to create a good RNG. (It was the first function that popped into my head.)
Consider the iteration

$$X_{n+1} = |100 \log(X_n) \bmod 1|, \quad n = 0, 1, 2, 3, \dots$$

where $x_0 = 0.1$. By "mod 1," I mean that only the digits to the right of the decimal point are used. The log function is the natural log. The Cliff algorithm is easily implemented in C using double precision numbers:

```
long num, n;
double x, xi;
x = 0.1 /* seed */
num = 1e6;
for (n=0; n<num; n++) {
    x = 100*log(x); /* natural log */
    xi = trunc(x);
    x = fabs(x-xi); /* absolute value */
    printf("Random number is: %.16f\n",x);
}
```

(If desired, you may add a test to check that the value for x never becomes 0. I did not find this necessary in my applications.)

In a good RNG, we would like the numbers produced to be spread evenly over the range of 0 to 1. For example, it would not be useful if many more numbers were produced between 0 and 0.5 as between 0.5 and 1. Figure 31.1 suggests that X_n are sufficiently uniformly distributed on the interval (0, 1) to be useful for a wide variety of applications. The Cliff RNG also has a theoretically infinite period; however, in practice the period is finite due to limited machine precision. I find that we can easily generate a million uniformly distributed numbers on an IBM RISC System/6000 in about 4 seconds, with no cycling of numbers. Obviously my goal here is not to create the most efficient RNG or one with a superlong period—but simply one that is easy to remember and adequate for many purposes.

To assess possible correlations between neighbor values in X_n, I used a highly sensitive graphical approach called the noise-sphere, which I first described in my book *Computers and the Imagination*. (In harmony with the general objectives of this chapter, my goal here is also to present simple, easy-to-remember, and easy-to-program approaches.) The noise-sphere is computed by mapping random number

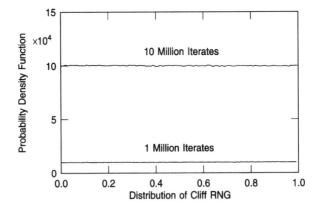

FIGURE 31.1.   Histogram with 100 intervals computed from 1 million iterates and 10 million iterates of the Cliff random number generator X_n, as described in the text for $X_0 = 0.1$.

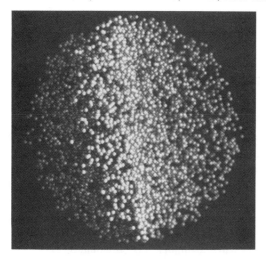

FIGURE 31.2.   Noise-sphere for the Cliff RNG. No particular unwanted correlations are evident.

triplets to the position of little balls in a spherical coordinate system, as described in the next section. High-quality RNGs do not show any particular features in the noise-sphere; low-quality ones yield tendrils and various forms of nonhomogeneity. Not only do all good RNGs I have tested pass the noise-sphere test, but also all famous bad generators with correlations fail it.

Figure 31.2 (discussed in more detail in the next section) indicates that the Cliff RNG produces essentially uncorrelated numbers, because no particular features are evident in the collection of little spheres. As a counterexample, examine the noise-sphere in Figure 31.3, which omits the 100 from the expression $100 \times \log(x)$ for the Cliff RNG. The nonindependence of the numbers is clearly seen. In the next section,

FIGURE 31.3.   Noise-sphere for a bad random number generator.

I describe the noise-sphere and further suggest its usefulness as a simple way to check the quality of random number generators.

Noise-Spheres

A sensitive graphical representation called a noise-sphere can be used to detect many "bad" random number generators with little training on the part of the observer. (Bad random number generators often show various tendrils when displayed using the noise-sphere representations.) As suggested twice before in this chapter, in modern science, random number generators have proven invaluable in simulating natural phenomena and in sampling data. It is therefore useful to build easy-to-use graphic tools that, at a glance, help to determine whether a random number generator being used is "bad" (that is, nonuniform and/or with nonindependence between various digits). Although a graphics supercomputer facilitates the particular representation described in the following, a simpler version would be easy to implement on a personal computer. I want to stimulate and encourage programmers, students, and teachers to explore this technique in a classroom setting.

To produce the noise-sphere figures, simply map the output of an RNG to the position of spheres in a spherical coordinate system. This is easily achieved with a computer program using random numbers $\{X_n, n = 0, 1, 2, 3, \ldots, N \ (0 < X_n < 1)\}$, where X_n, X_{n+1}, and X_{n+2} are converted to θ, φ, and r, respectively. For each Cartesian triplet, an (r, θ, φ) is generated, and these coordinates position the individual spheres. The data points may be scaled to obtain the full range in spherical coordinates:

$$2\pi X_n \rightarrow \theta$$
$$\pi X_{n+1} \rightarrow \varphi$$
$$\sqrt{X_{n+2}} \rightarrow r$$

The square root for X_{n+2} serves to spread the ball density though the spherical space and to suppress the tight packing for small r. (As a reminder, to convert from θ, φ, r to x, y, z use a spherical transformation: $x = r \sin\theta \cos\varphi$, $y = r \sin\theta \sin\varphi$, $z = r \cos\theta$.) Each successive triplet overlaps, so that each point has the opportunity to become an r, θ, or φ. (The advantage over a Cartesian system is discussed later in this chapter.)

The resultant pictures can be used to represent different kinds of noise distributions or experimental data. As already mentioned, no particular correlations in the ball positions are visually evident with good random number generators. For surprising results when this approach is applied to a BASIC random number generator, see my book *Computers and the Imagination*. In particular, the method has been effective in showing that random number generators prior to release 3.0 of BASIC have subtle problems.

Although such representations are fairly effective in black and white, I have found it useful to map the random number triplet values to a mixture of red, green, and blue intensities to help the eye see correlated structures. (In other words, for a particular sphere, a value of red, green, and blue from 0 to 1 is determined by the value of X_n, X_{n+1}, and X_{n+2}.) I use an IBM Power Visualization System to render and rotate the noise-spheres; however, interested users without access to graphics supercomputers can render their data simply by plotting a scattering of dots and projecting them on a plane. Most important, I have found that the spherical coordinate transformation used to create the noise-sphere allows the user to see trends much easier, and with much fewer trial rotations, than an analogous representation that maps the random number triplets to three orthogonal coordinates. Because the noise-sphere approach is sensitive to small deviations from randomness, it may have value in helping researchers find patterns in complicated data such as genetic sequences and acoustical waveforms.

Some readers may wonder why I should consider the noise-sphere approach over traditional brute force statistical calculations. One reason is that this graphic description requires little training on the part of users, and the software is simple enough to allow users to quickly implement the basic idea on a personal computer. The graphics description provides a qualitative awareness of complex data trends and may subsequently guide the researcher in applying more traditional statistical calculations. Also, fairly detailed comparisons between sets of "random" data are useful and can be achieved by a variety of brute force computations, but sometimes at a cost of the loss of an intuitive feeling for the structures. When we just look at a page of numbers, differences between the data's statistics may obscure the similarities. The approach described here provides a way of simply summarizing comparisons between random data sets and capitalizes on the feature integration abilities of the human visual system to see trends.

Explore Extraterrestrial Lands

In this section, I extend our discussion of simple RNGs to new functions and introduce readers to a fascinating, nonperiodic, near-Gaussian random number generator that was studied in the past by Dr. James Collins (Boston University) and colleagues. (A Gaussian distribution is a bell-shaped curve, symmetrical about the average value.) I find that this near-Gaussian RNG is useful for a variety of purposes, including the graphical generation of extraterrestrial terrain (Figure 31.4), and I am interested in hearing from readers who use these methods for other applications. The described approach may be utilized in laboratory and classroom demonstrations, and has the potential for producing a wide range of fanciful terrain structures for movie special effects houses.

The Bizarre Logistic Equation

To produce Figure 31.4, I have used an approach based on the famous logistic equation that is easily implemented by beginning programmers and computer graphics aficionados. The logistic equation or map is given by the expression

$$x_{n+1} = 4\lambda x_n(1 - x_n)$$

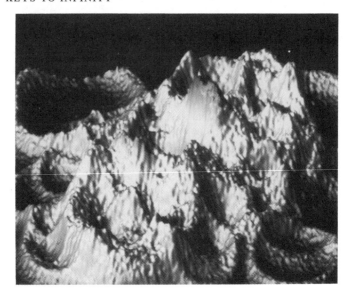

FIGURE 31.4.   Terrain generated using the radialized logit transform on the logistic variable.

where $0 < \lambda < 1$. This equation was introduced in 1845 by the Belgian sociologist and mathematician Pierre-Francois Verhulst to model the growth of populations limited by finite resources. The equation has since been used to model such diverse phenomena as fluid turbulence and the fluctuation of economic prices.[1] As the parameter λ is increased, the iterates x_n exhibit a cascade of period-doubling bifurcations followed by chaos.

If we use $\lambda = 1$, then x_n are "random" numbers on the interval $(0, 1)$ with a U-shaped probability distribution P_x given by:[2,3,4]

$$P_x = \frac{1}{\pi\sqrt{x(1-x)}}$$

Figure 31.5 shows a histogram with the expected clustering near $x = 0$ and $x = 1$ for $x_0 = 0.1$.

In 1964, Ulam and von Neumann defined a new variable y_n as

$$y_n = (2/\pi)\sin^{-1}(\sqrt{x_n})$$

This "Ulamized" variable y_n has potential as a random number generator because the numbers produced are uniformly distributed on the interval $(0, 1)$ (see Figure 31.5).

Note that although the Ulamized variables are distributed uniformly on the interval $(0, 1)$, unfortunately the variables do have correlations. This is also true for the logistic variables, as evidenced by the noise-spheres in Figures 31.6 and 31.7. Therefore, users of this approach will have to sample the generated variables at some suitably large interval to eliminate correlations. I recommend users skip iterates, taking every tenth iterate, at which point no correlations are visually apparent (Figure 31.8).

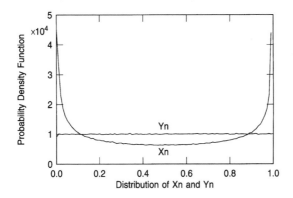

FIGURE 31.5.   Histogram with 100 intervals computed from 1 million iterates of the logistic equation x_n, and the Ulamized logistic equation y_n, for $x_0 = 0.1$.

Interestingly, the noise-spheres are more useful than traditional correlation functions when applied to many kinds of random data. For example, Grossman and Thomae have computed correlation functions for the logistic formula, and the functions appear to indicate vanishingly small (close to 0) correlations for lags greater than 1.[5] I have confirmed this (see Figures 31.9 and 31.10) by computing the same kind of autocorrelation function for the logistic variables (and for other RNGs):

$$R_x(p) = \frac{1}{n-p} \sum_{q=1}^{n-p} X_q X_{q+p} \quad p = 0, 1, 2, \ldots, m$$

where n is the number data points.

FIGURE 31.6.   Noise-sphere for logistic variables, lag = 1. Here no points are skipped. By skipping points (see Figures 31.7 and 31.8), correlations are reduced.

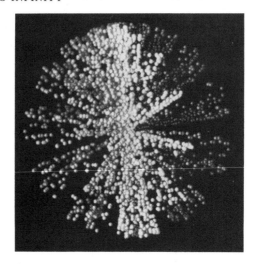

FIGURE 31.7.   Noise-sphere for logistic variables, lag = 5. (Every fifth point is used. The others are discarded.)

The autocorrelation function for random data describes the general dependencies of the values of the data at one point in the sequence on the values at another place. It is a measure of the predictability of a fluctuating function $x(t)$. If the value of the function $x(t + \tau)$ can be predicted with a good expectation of success when the value of $x(t)$ is known, then the product $x(t)x(t + \tau)$ does not become zero when averaged over t. Grossman and Thomae examine such autocorrelation functions for the logistic formula, note the immediate drop to zero, and say that for $\lambda = 1$ "all the iterates are ergodic, that is [the logistic formula] generates pure chaos." I have found that many graphics specialists have taken this to suggest that the logistic formula

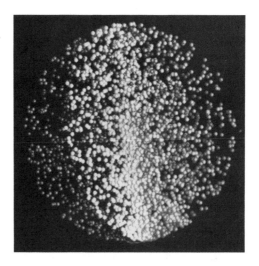

FIGURE 31.8.   Noise-sphere for logistic variables, lag = 10.

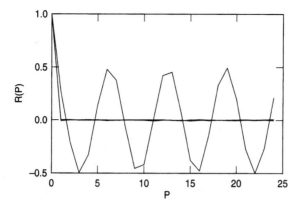

FIGURE 31.9. Autocorrelation function for a sine wave (the rippling function), and for the logistic, Cliff, and C-language random number generator (all which immediately drop to values close to zero).

produces uncorrelated values. (Note that the noise-sphere allows humans to perceive correlations even beyond a lag of 5.) In other words, individuals sometimes confuse the concept of ergodic orbits of dynamical systems with the concept of an independent, identically distributed sequence of numbers that ideally is produced by an RNG. Sequences of numbers produced by function iteration ($X_{n+1} = f(X_n)$) can almost never be independent, identically distributed sequences, and often are highly correlated as evidenced in this chapter. One standard way of revealing this correlation is simply to plot X_n versus X_{n+1} (see, for example the article by T. Richards).[6] The noise-spheres do a similar job but show correlations among three successive values.

Figure 31.9 shows the autocorrelation function for a sine wave (the rippling function), and for the logistic, Cliff, and C-language random number generator (all

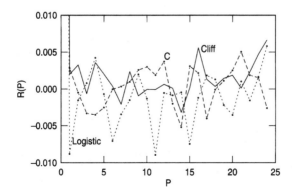

FIGURE 31.10. Magnification of Figure 31.9.

which immediately drop to values close to zero). As can be seen, except for the sine wave, the others all have a vanishingly small correlation value for lags greater than 1. Even when the autocorrelation functions for the various random number generators are greatly magnified (Figure 31.10), not much useful information can be gleaned from the noisy plots. The noise-spheres are a much better way to detect correlations within noisy data sets. For a sampling of brute force statistical tests that are sensitive to correlations, see *The Art of Computer Programming*.[7]

The Logit Transformation

For terrain generation, student experiments, and possible inclusion in statistical software packages, I have found the transformation given by

$$z = \ln\left(\frac{x}{1-x}\right) \tag{1}$$

to be quite useful. (In statistics this is known as the logit.) The logit transform may operate *directly* on the logistic variable, and Collins et al. (refer to note 2) have shown that this transformation generates a sequence of random numbers with a near-Gaussian distribution:

$$P_z = \frac{1}{\pi}\left(\frac{e^{x/2}}{1+e^x}\right) = \frac{1}{\pi(e^{x/2}+e^{-x/2})}$$

(Students are urged to compare this to a Gaussian distribution and observe the remarkable similarity.) Figure 31.11 shows an experimentally derived histogram for the logit transform of the logistic variable.

If readers are interested in using the preceding for 3-D extraterrestrial terrain generation, convert the z_n distribution to a radial one in polar coordinates, and use the Ulamized variables y_n as angles:

$$v = z_n \alpha \sin(2\pi y_n) + cx$$
$$w = z_n \alpha \cos(2\pi y_n) + cy \tag{2}$$

For each d(v, w) pair generated, a two-dimensional $P(v, w)$ array is incremented by 1 at location (v, w). (Care should be taken to iterate the logistic formula a few times before using it to compute both y_n and z_n, to eliminate unwanted correlations.) The resultant $P(v, w)$ distribution, after many iterations, resembles a circular ridge centered at (cx, cy) with a near-Gaussian cross-section. The variable α controls the diameter of the crater. The landscape in this chapter was created by computing a 512-by-512 $P(v, w)$ array for 50 Ulamized values of α, cx, and cy. In other words, a random (α, cx, cy) triple is selected using Ulamized logistic numbers, and then equation 2 is iterated N times ($N \leq 10^6$) and a $P(v, w)$ histogram is created. A new triple is chosen, and the process is repeated—each time "laying down" a radial, near-Gaussian function as numbers accumulate in the $P(v, w)$ histogram. For more examples of

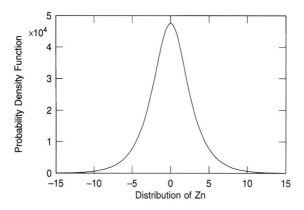

FIGURE 31.11. Histogram with 100 intervals computed from 1 million iterates of the logit transform of the logistic variable for $x_0 = 0.1$.

terrain produced by this method, more computational details, and further analysis, see my *IEEE* article.[8]

In recent years, the generation of both realistic and fanciful terrain has been of substantial interest to computer graphics specialists, motion picture and special effects artists, geologists, and even mathematicians. I enjoy the logit method because it produces a wealth of lunar-like structures not produced by standard fractal methods, and also the method is conceptually simple when compared to Fourier transform and midpoint displacement methods. It is easily implemented and understood by students and programmers. There are no Fourier transforms, no erosion models, no convolutions, no complicated vertex bookkeeping. From a pedagogical standpoint, it therefore makes a wonderful demonstration of not only how to build complex terrains but also how the relationships among random number generators, probability theory, and chaotic dynamics are intimately intertwined. From a practical standpoint, the logit transform operating on the logistic variables produces a distribution with a slightly more gradual decay to zero than does a Gaussian. This may be useful in certain applications; for example, for the terrain it means that the crater meets the ground in a more gradual way, which I find aesthetically pleasing.

Pretty Good Random Numbers

Obviously there are areas in physics and numerical analysis where "pretty good" RNGs will not be suitable, and various past papers will point readers to more sophisticated algorithms. On the other hand, I feel there is a need for simple, easy-to-remember RNGs such as the ones I have presented for various applied and pedagogical applications. Certainly the myth suggesting that adequate, simple RNGs are very difficult to create is one to explore. Interestingly, some researchers have suggested that RNGs should be able to give the same random numbers when run on different computers and when implemented in different languages. Unfortunately, as pointed out by Colonna, rounding errors often make this quite difficult, especially for generators based on function iterations of the form $X_{n+1} = f(X_n)$.[9]

Despite the strong correlations between successive numbers—when numbers are not skipped—there are several interesting advantages of the logit approach to producing near-Gaussian "random" numbers. First, these near-Gaussian numbers are easily incorporated into software for stochastic modelling. Second, like the Cliff RNG, the logistic-based number generators theoretically have an infinite period—although in practice the period is finite due to the limited precision of computers. This limitation can be easily overcome by restarting the generator with a new x_0 every billion or so iterations. (Would it help to change λ very slightly every billion or so iterations?) For the terrain in this chapter, x_0 and λ were fixed at constant values.

James Collins and colleagues have explored the use of the logit transform for a variety of statistical purposes, and they point out an interesting theoretical advantage that the nth number generated by the logistic equation can be determined directly without iteration:

$$x_n = \sin^2[2^n \sin^{-1}(\sqrt{x_0})]$$

where x_0 is some initial value. The values of x_n generated in this manner can then be transformed by the logit given in equation 1.

Interest in random number generators continues to grow as evidenced by a recent explosion of papers on exciting new random generators that are not based on the extensively studied multiplicative congruential types. For example, much longer "defect-free" sequences have been proposed that are based on chaotic dynamical systems. (See, for example, the RANLUX (luxury random numbers) algorithm, described in an article by F. James, which has an amazing period of 10^{170}.)[10] Various references to these new kinds of RNGs are included in "For Further Reading" at the end of this book.

A chapter such as this can only be viewed as introductory; however, it is hoped that the discussion will stimulate future interest in the graphic representation of simple random number generators. In laboratory exercises for courses such as statistical mechanics, thermodynamics, and even computer programming, it is unlikely that undergraduate students will be required to write their own RNGs, but I feel examples such as the ones given in this chapter will be of great instructional value to many students and scientists involved with computers. For information on other graphical methods for detecting patterns in random numbers, such as the use of computer-generated cartoon faces and Truchet tiles, see my previous books. For a fanciful treatment of RNGs in an alien society, see *Chaos in Wonderland*. For additional information and methods, including applications to cryptography, see Schroeder.

Flipping a Quantum Mechanical Coin

Physicists have always considered the processes by which light is emitted from an excited atom to be random. For example, we cannot predict when an electron will fall from an excited state to a lower energy state thereby causing a photon of light to be ejected from an atom. Thomas Erber of the Illinois Institute of Technology recently tested the randomness of such a process by examining light emitted by a single mercury ion. Applying a number of different techniques used both to assess the quality of random number generators and to search for patterns in encrypted data, Erber and co-workers found no patterns in 20,000 numbers representing the time intervals between light emission. In the July 16, 1994, *Science News*, Ivars Peterson notes that the light emission of a single ion may provide an infinite source of "cryptographically invulnerable" random numbers. Would the noise-spheres discussed in this chapter be useful for studying random numbers produced by electron transitions?

Randomness and Infinity

Is it possible for a computer algorithm to generate an infinite string of random numbers? Could the infinite digits of π be used as a very good source of random (that is, "normal") numbers for simulations? (A normal number is defined as a number in which every possible block of digits is equally likley to occur.) If so, why are they not used to generate a normal sequence of numbers? One advantage of using π is that researchers and statisticians could easily repeat one another's computational experiments. A disadvantage is the difficulty of computing or storing the digits of π. (As mentioned in other chapters, tables already exist for more than 2 billion digits of π.)

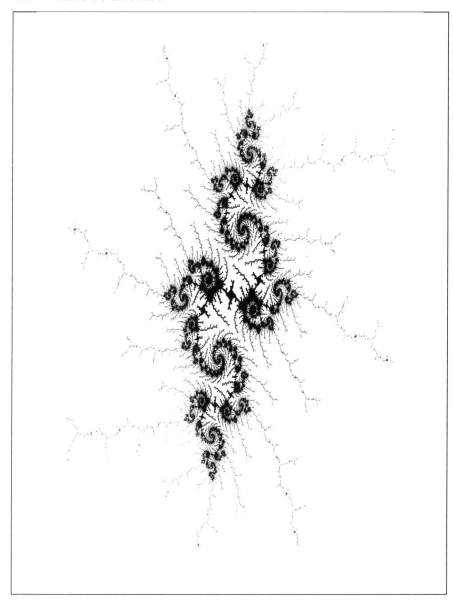

"And seize the tiptoe lightnings, ere they fly. . . ." (E. Darwin, 1789)
This lightning fractal was produced by methods similar to those discussed in chapters 6 and 23. Could the tendrils be used to create a random number generator?

Program Code

Chapter 1 Too Many Threes

C Code

```
/* Compute Proportion of Numbers That Contain the Digit 3 */
/* See text for a challenge to create a related, more
   interesting program. */

#include <math.h>
#include <stdio.h>
int n;
float proportion;
main()
{
  for (n=0; n <= 100; n++) {
     proportion = pow((9.0/10.0),n);
     printf("%d %f %f n",n,proportion,1.-proportion);
  }

}
```

BASIC Code

```
10 REM Compute Proportion of Numbers That Contain the Digit 3
15 REM See text for a challenge to create a related, more
17 REM interesting program.
20 FOR N=0 TO 100
25    REM Some BASIC languages use ¬ instead of **
30    P = (9.0/10.0)**N
40    PRINT N, P, 1-P
50 NEXT N
60 END
```

Chapter 2 Ladders to Heaven

C Code

```
/* Approximate the ratio of weight in space to weight on Earth */
#include <math.h>
```

```c
#include <stdio.h>

main()

{
    float step, Rm, W, Me, Mm, Re, D, d1, d2;

    Me = 5.97e24; /* Mass of Earth kg */
    Mm = 7.35e22; /* Mass of moon kg */
    Re = 6371;    /* Mean radius of Earth km */
    D  = 384401;  /* Mean Earth-moon distance km */
    Rm = 1738;    /* Mean radius of moon km */
    step = (D - (Re+Rm))/100.; /* sample graph using 100 points */
    /* d1 is distance from center of Earth to point out in space */
    for (d1 = Re; d1 <= (D-Rm); d1 = d1 + step) {
        d2 = D - d1;
        W = (Me/(d1*d1) - Mm/(d2*d2)) * (Re*Re)/Me;
        printf("Ratio is: %f, Distance is: %f n",W,d1);
    }
}
```

BASIC Code

```
100 REM APPROXIMATE THE RATIO OF WEIGHT IN SPACE TO WEIGHT ON EARTH
110 REM MASS OF EARTH, KG
120 E = 5.97E24
130 REM MASS OF MOON, KG
140 M = 7.35E22
150 REM MEAN RADIUS OF EARTH, KM
160 R = 6371
170 REM MEAN EARTH-MOON DISTANCE, KM
180 D = 384401
190 REM MEAN RADIUS OF MOON, KM
200 T = 1738
210 REM SAMPLE GRAPH USING 100 DISTANCE POINTS
220 S = (D - (R+T))/100
230 G IS DISTANCE FROM CENTER OF EARTH TO POINT OUT IN SPACE
330 FOR G = R TO (D-T) STEP S
340    F = D - G
350    W = (E/(G*G) - M/(F*F)) * (R*R)/E
360    PRINT "RATIO IS "; W; "DISTANCE IS"; G
370 NEXT G
380 END
```

Chapter 5 Grid of the Gods

C Code

```c
/* Compute chords of circle */
#include <math.h>
#include <stdio.h>
 int i,count1,count2;
 float pi,dist,theta,r,x,y,x2,y2,ratio;
 FILE *fp,*fp2;
```

```
main()
{
    r = 1;   pi = 3.1415926;
    fp = fopen("grid1.out","w");
    fp2 = fopen("grid2.out","w");
    srand(1234567);  /* random number seed */
    count1=0; count2=0;
    for (i=0; i<1000; i++) {
        /* select random point on circle */
        theta = 2*pi*((float)rand()/32767.);
        x = r*cos(theta);
        y = r*sin(theta);
        /* select another random point on circle */
        theta = 2*pi*((float)rand()/32767.);
        x2 = r*cos(theta);
        y2 = r*sin(theta);
        dist = sqrt((x-x2)*(x-x2) + (y-y2)*(y-y2));
        if (dist<r) count1++;
        count2++;
        ratio=(float)count1/(float)count2;
        if ((i%10)==0) fprintf(fp,"%d %f\n",i,ratio);
        if ((i%10)==0) printf("%d %f\n",i,ratio);
        if (i<100) fprintf(fp2,"%f %f\n",x,y);
        if (i<100) fprintf(fp2,"%f %f\n",x2,y2);
    }
    printf("Final: %d %d %f\n", count1, count2, ratio);
}
```

BASIC Code

```
10 REM Compute chords of circle
20 R = 1.0
30 P = 3.1415926
40 C1 = 0
50 C2 = 0
60 FOR I=0 TO 100
70     REM select random point on circle
80     T = 2.0*P*RND
90     X = R*COS(T)
100    Y = R*SIN(T)
110    REM select another random point on circle
120    T = 2.0*P*RND
130    X2 = R*COS(T)
140    Y2 = R*SIN(T)
150    D = SQR((X-X2)*(X-X2) + (Y-Y2)*(Y-Y2))
160    IF D<R THEN C1=C1+1
170    C2 = C2+1
190    REM Compute Ratio
200    S=C1/C2
210    PRINT "Points on Circle: "; X;Y;X2;Y2
220    PRINT
230    PRINT I; D; "Ratio: "; S
240 NEXT I
250 END
```

Chapter 8 Infinite Chess

C Code

```
/* Solve bishop problem by printing values in the table */
#include <math.h>
#include <stdio.h>
 int n, maxbishop, totbishop, distinct;
 float ratio;
 FILE *fp;
main()
{
   for (n=2; n<=20; n++) {
      maxbishop = 2*n-2;
      totbishop = pow(2,n);
      if ((n%2)==0) distinct = pow(2,(n-4)/2)*(pow(2,(n-2)/2)+1);
      else distinct = pow(2,(n-3)/2)*(pow(2,(n-3)/2)+1);
      ratio = (float)totbishop/(float)distinct;
      printf("%d %d %d %d %f\n",n,maxbishop,totbishop,distinct,ratio);
   }
}
```

BASIC Code

```
10 REM Solve bishop problem by printing values in table
20 FOR N=2 TO 20 STEP 2
30    REM M is maximum number of nonattacking bishops
40    M = 2*N-2
50    REM T is total number of bishop arrangements
60    T=2**N
70    REM D is the number of distinct positions
80    D = 2**((N-4)/2)*(2**((N-2)/2)+1)
90    REM Insert next statement for odd numbers:
100   REM D = 2**((N-3)/2)*(2**((N-3)/2)+1)
110   REM Ratio:
120   R = T/D
130   PRINT N;M;T;D;R
140 NEXT N
150 END
```

Chapter 9 The Loom of Creation

C Code

```
/* Compute chords of circle */
#include <math.h>
#include <stdio.h>
 int i;
 float pi,theta,r,x,y,x2,y2;
 FILE *fp;
main()
{
   r = 1;  pi = 3.1415926;
```

```
/* create ranunculoid */
a = 6; b = 6;
/* create cardioid */
a = 2; b = 2;
/* create nephroid */
a = 3; b = 3;
/* create amazingoid */
a = 100; b = 100;
/* create fishtailoid */
a = 2; b = 3;
for (i=0; i<=360; i=i+1) {
    theta = (float)i*pi/180.;
    x = r*cos(theta);
    y = r*sin(theta);
    /* Select another point on circle */
    x2 = r*cos(a*theta);
    y2 = r*sin(b*theta);
    /* Print endpoints of each line */
    printf("%f %f\n",x,y);
    printf("%f %f\n",x2,y2);
    }
}
```

BASIC Code

```
10 REM Compute ranunculoids
20 R = 1
25 REM Ranunucloid parameters:
26 A = 6
27 B = 6
30 P = 3.1415926
40 FOR I = 0 TO 360
50     T = I*P/180.0
60     X = R*COS(T)
70     Y = R*SIN(T)
80     REM select another point on circle
90     X2 = R*COS(A*T)
100    Y2 = R*SIN(B*T)
110    PRINT "Chord on Circle: "; X;Y;" TO ";X2;Y2
120 NEXT I
130 END
```

Chapter 10 Slides in Hell

C Code

```
/* Simulate launching people down slide */
#include <math.h>
#include <stdio.h>
 int i, j, flag, bin{20000}, count;
 float x;
 FILE *fp;
main()
{
```

```
/* intialize bins */
for (i=0; i<20000; i++) bin{i}=0;
srand(123456789); /* random number gen seed */
count=0; i=0;
while (count<10000) {
   i++;          /* launch person down slide */
   flag=0;
   for (j=0; j<10; j++) {
      /* generate random number between 0 and 1 */
      x = ((float)rand()/32767.);
      if (x>0.5) {flag=1; break;} /* person fell through hole */
   }
   /* Include the following line if you want print to the screen */
   /* if (flag==0) {printf("Made it in %d tries\n",i); } */
   if (flag==0) {
         /* person has made it down the slide */
         count++;
         bin{i}=bin{i}+1;
         i=0;  /* reset */
         if ((count%1000)==0) printf("%d\n",count);
   }
} /*while */
/* Dump output file showing distribution */
fp = fopen("slide.out", "w");
for (j=0; j<4000; j++) {
      fprintf(fp,"%d %d\n", j,bin{j});
}
}
```

BASIC Code

```
10 REM Simulate launching people down slide
20 DIM B(20000)
30 REM Initialize bins for histogram
40 FOR I=1 TO 20000
50    B(I)=0
60 NEXT I
70 C=0
80 I=0
110 REM Launch person down slide
120 I=I+1
130 F=0
140 FOR J=1 TO 10
150    REM Generate random number between 0 and 1
160    X=RND
170    REM X>0.5 means person fell through a hole
190    IF X>0.5 THEN GOTO 110
200 NEXT J
210 REM Include the following line if you want print to the screen
220 PRINT "Made it in ";I;"tries."
240 REM Person has made it down the slide
250 C=C+1
260 B(I)=B(I)+1
270 REM RESET
280 I=0
```

```
290 IF C=1001 THEN PRINT "More than 1000 people have made it down."
300 IF C<10000 THEN GOTO 110
310 REM Print distribution
320 FOR J=1 TO 4000
330    PRINT J; B(J)
340 NEXT J
350 END
```

Chapter 11 Alien Abduction Algebra

C Code

```c
/* Compute number and sex of humans */
#include <math.h>
#include <stdio.h>
main()
{
    int i, j, countone, counttwo;
    double k, ratio;
    FILE *fp;
    fp=fopen("output.file","w");

    countone=0; counttwo=0;
    k=(sqrt(5.0)+1.0)/2.0;
    for (i=1; i<=10; i++) {
        j=trunc(k*i)-trunc(k*(i-1));
        printf("Term %d is %d\n",i,j);
        if (j==1) countone++;
        if (j==2) counttwo++;
        if (i>1) {
          ratio=(float)countone/(float)counttwo;
          printf("Cumulative number of males %d n",countone);
          printf("Cumulative number of males %d n",counttwo);
          printf("The ratio is %f n",ratio);
        }
        fprintf(fp,"%d %d %d\n",i,countone,counttwo);
          /* for graph */
    } /* for */
}
```

BASIC Code

```basic
100 REM Compute number and sex of humans
110 M=0
120 F=0
121 REM S5 is a high-precision value for SQR(5)
122 S5=2.23606797749978969640917
125 K=(S5+1.0)/2.0
126 PRINT K
130 FOR I=1 TO 50
140    J=INT(K*I) - INT(K*(I-1))
150    PRINT "TERM"; I; " IS "; J;
155    PRINT " "
160    IF J=1 THEN M=M+1
```

```
170    IF J=2 THEN F=F+1
180    IF I>1 THEN GOSUB 300
190 NEXT I
200 STOP
300 R=M/F
310 PRINT "Cumulative number of males "; M
320 PRINT "Cumulative number of females "; F
330 PRINT "The ratio is "; R
340 RETURN
500 END
```

Chapter 12 The Levianthan Number

C Code

```c
/* Compare Stirling with standard factorial values */
#include <math.h>
#include <stdio.h>
int i, n;
float pi=3.1415926,fact,f[100],s[100],diff;
main()
{
  /* Compute Stirling values */
  for (n=0; n<30; n++) {
      s[n]=sqrt(2.*pi)*exp(-n)*pow(n,n+0.5);
  }
  /* compute factorial */
  f[1]=1;
  for (n=1; n<30; n++) {
      f[n+1]=n*f[n];
  }
  /* Compute percentage difference */
  for (n=1; n<30;n++) {
    diff=(f[n+1]-s[n])/f[n+1];
    printf("%d %f %f %f\n",n,f[n+1],s[n],diff);
  }

}
```

BASIC Code

```
10 REM Compute Stirling and factorial values
20 DIM F(100), S(100)
30 P=3.1415926
35 REM Compute factorial using Stirling's formula
37 FOR N=1 TO 30
40     S(N)=SQR(2.*P)*EXP(-N)*N**(N+0.5)
50 NEXT N
55 REM Compute factorial
60 F(1)=1
70 FOR N=1 TO 30
80     F(N+1)=N*F(N)
90 NEXT N
```

```
95 REM Compute factorial
100 FOR N=1 TO 30
110    D=(F(N+1)-S(N))/F(N+1)
120    PRINT N;F(N+1);S(N);D;
130    PRINT
140 NEXT N
150 END
```

Chapter 13 Welcome to Worm World

C Code

```c
/* Generate and explore Worm Worlds */
#include <math.h>
#include <stdio.h>

int l, p[200], icur, jcur, m, n, i, j, k, x[100][100];
int flag, path_len;
main()
{
 /* random number gen seed */
 srand(131313131);
 /* generate array of numbers */
 k=0;
 for (i=0; i<8; i++) {
   for (j=0; j<8; j++) {
     k++;
     /* generate random numbers (0 - 24) */
     x[i][j] = ((float)rand()/32767.)*25.;
     printf("%2d ",x[i][j]);
     if (((k)%8)==0) printf(" n");
   } /* for */
  } /* for */
/* for each cell, scan array for longest path */
 printf("Paths: n");
 l=0;
 for (i=0; i<8; i++) {
   for (j=0; j<8; j++) {
     l++;
     path_len = 1;
     for (k=0; k<100; k++) p[k]=0;
       p[0]=x[i][j]; /* first point in path */
       icur=i;  jcur=j;
       for (n=0; n<64; n++)  {
         /* check neighborhood */
         jcur=jcur+1; scan_path();
         if (flag==1) {jcur=jcur-2; scan_path();
         if (flag==1) {icur=icur+1; jcur=jcur+1; scan_path();
         if(flag==1) {icur=icur-2; scan_path();
         if(flag==1) {write_path(); n=100; /* break out */
                    } } } }  /* if's */
       } /* for n */
   } /* for j */
```

```
    } /* for i */
} /* main */


/* scan path for "reused" numbers */
scan_path()
{
        /* flag=1 means unsuccessful */
        flag=0;
        if((icur<0) || (jcur<0) || (icur>7) || (jcur>7)) flag=1;
            for (k=0; k<path_len; k++) {
                if(x[icur][jcur]==p[k]) flag=1;
            }
        if(flag==0) {p[path_len]=x[icur][jcur]; path_len++;}

}
/* Write out "longest" path for each starting cell  */
write_path()
{printf("(%d %d) ",l,path_len);
    for (k=0; k<path_len; k++) printf("%d ",p[k]); printf("\n"); }
```

BASIC Code

```
100 REM /* Generate and explore Worm Worlds */
200 DIM P(100), X(100,100)
300 REM Generate array of numbers
400 K=0
500 FOR I=1 TO 8
600     FOR J=1 TO 8
700         REM Generate random numbers (1 - 25)
800         X(I,J)=INT(RND*25)
900     NEXT J
1000    REM Print board
1200    PRINT X(I,1); X(I,2);X(I,3);X(I,4);X(I,5); X(I,6); X(I,7); X(I,8)
1300 NEXT I
1400 REM For each cell, scan array for longest path
1500 PRINT "PATHS:"
1600 L=0
1700 FOR I=1 TO 8
1800    FOR J=1 TO 8
1900        L=L+1
2000        REM P1 is path length
2100        P1=1
2200        FOR K=1 TO 100
2300            P(K) = 0
3400        NEXT K
3500        REM First Point in Path
3600        P(1)=X(I,J)
3700        I1=I
3800        J1=J
3900        FOR N=1 TO 65
4000            REM Check neighborhood
4100            J1=J1+1
4200            GOSUB 10000
4300            IF F=1 THEN J1=J1-2
```

```
4600              IF F=1 THEN GOSUB 10000
4700              IF F=1 THEN I1=I1+1
4800              IF F=1 THEN J1=J1+1
4900              IF F=1 THEN GOSUB 10000
5000              IF F=1 THEN I1=I1-2
5100              IF F=1 THEN GOSUB 10000
5200              IF F=1 THEN GOSUB 20000
5300              REM Now break out of For Loop
5400              IF F=1 THEN GOTO 5600
5500          NEXT N
5600        NEXT J
5700    NEXT I
5800    STOP
10000   REM Scan path for "reused" numbers
10100    REM FLAG=1 means unsuccessful
10200    F=0
10300    IF I1<1 THEN F=1
10400    IF J1<1 THEN F=1
10500    IF I1>8 THEN F=1
10600    IF J1>8 THEN F=1
10650    IF F=1 THEN GOTO 11200
10700    FOR K=1 TO P1
10800        IF X(I1,J1)=P(K) THEN F=1
10900    NEXT K
11000    IF F=0 THEN P1=P1+1
11100    IF F=0 THEN P(P1)=X(I1,J1)
11200    RETURN
20000   REM Write out "longest" path for each starting cell
20100    PRINT "CELL, PATH LENGTH";L;P1
20200    FOR K=1 TO P1
20300        PRINT P(K)
20400    NEXT K
20500    RETURN
20600    END
```

Chapter 14 Fractal Milkshakes and Infinite Archery

C Code

```c
/* Compute Ford froth */
#include <math.h>
#include <stdio.h>

main()
{
    float rad, cx, cy, s, x, y, z, k, h;
    /* Scan 15 values for h and k */
    for (h=1; h<=15; h=h+1) {
        for (k= 1; k<=15; k=k+1) {
            rad=1./(2.*k*k);
            cx=h/k;
            cy=rad;
            /* Print out points suitable for drawing each circle */
            /* The step size of .1 controls the circle resolution */
```

```
            for (s=0; s<=6.28; s=s+.1){
                x=rad*cos(s)+cx;
                y=rad*sin(s)+cy;
                printf("%f %f\n",x,y);
            }
        /* Next compute another circle */
        }
    }
}
```

BASIC Code

```
10  REM Compute Ford froth */
20  REM Scan 15 values for h and k
30  FOR H=1 TO 15
40          FOR K=1 TO 15
50                  R=1./(2.*K*K)
60                  C1=H/K
70                  C2=R
90                  REM Dump points for drawing each circle
90                  REM The step value controls the circle fineness
100                 FOR S=0 TO 6.28 STEP .1
110                         X=R*COS(S)+C1
120                         Y=R*SIN(S)+C2
130                         PRINT X,Y
140                 NEXT S
150                 REM Next compute another circle
160         NEXT K
170 NEXT H
180 END
```

Chapter 15 Creating Life Using the Cancer Game

C Code

```
/* Compute Cancer Game Sequence */
#include <math.h>
#include <stdio.h>
#define max(a,b) ((a)>(b)?(a):(b))

main()
{
    /* last, 2nd to last, and current string */
    int l1[5000], l2[5000], cur[5000];
    /* size of last, 2nd to last, current */
    int sl1, sl2, scur;
    int i, j, temp, gen, numgen, maxlen;

    /* G=1, C=2, A=3, T=4 */
    /* initial two seeds */
    l2[0]=1; sl1=1; /* G; length 1 */
    l1[0]=2; sl2=1; /* C; length 1 */
    numgen=7; /* how many generations */
```

```
maxlen=0;
printf("G nC n");

for (gen=0; gen<numgen; gen++) {
    /* MUT Enzyme reads second-to-last generation */
    j=0; /* counter for string length */
    for (i=0; i<sl2; i++) {
        cur[j]=l2[i];
        /* Mutate: */
        if(cur[j]==1) temp=3;
        if(cur[j]==2) temp=4;
        if(cur[j]==3) temp=1;
        if(cur[j]==4) temp=2;
        cur[j]=temp;
        j++;
    }

    /* COP Enzyme reads last generation */

    for (i=0; i<sl1; i++) {
        cur[j]=l1[i];
        j++;
    } /* endfor */
    /* Print current line */
    for (i=0; i<j; i++) {
        if(cur[i]==1) printf("G");
        if(cur[i]==2) printf("C");
        if(cur[i]==3) printf("A");
        if(cur[i]==4) printf("T");
    }
    printf(" Length: %d\n",j);
    maxlen=max(j,maxlen);

    /* Compute Cancer Game Sequence (CONTINUED) */
    /* CLEAV splits sequence at T-T */
    scur=j;
    for (i=0; i<scur; i++) {
        if((cur[i]==4)&&(cur[i+1]==4))
        {
            scur=i+1; /* new length */
            printf("Cleavage has occurred n");
            for (i=0; i<scur; i++) {
                if(cur[i]==1) printf("G");
                if(cur[i]==2) printf("C");
                if(cur[i]==3) printf("A");
                if(cur[i]==4) printf("T");
            }
            printf("\n");
        } /* endif */
    } /* endfor */

/* Update by swapping generations*/
for (i=0; i<sl1; i++) l2[i]=l1[i];
sl2=sl1;
```

```
    for (i=0; i<scur; i++) ll[i]=cur[i];
    sll=scur;
} /* end For generations */
printf("Maximum string length found is: %d\n",maxlen);
} /* end program */
```

BASIC Code

```
100  REM Compute Cancer Game Sequence
200  REM Last, 2nd to last, and current string
300  DIM L(5000), M(5000), C(5000)
400  REM Size of last, 2nd to last, Current
500  REM S1, S2, S3
600  REM G=1, C=2, A=3, T=4
700  REM Initial two seeds:
750  REM G, length 1
800  M(1)=1
850  S1=1
860  REM C, length 1
900  L(1)=2
950  S2=1
1000 REM How many generations
1100 N=7
1200 REM Maximum length
1300 P=0
1400 PRINT "G"
1500 PRINT "C"
1600 FOR G=0 TO N
1700    REM MUT Enzyme reads second-to-last generation
1800    REM Counter for string length
1900    J=1
2000    FOR I=1 TO S2
2100      C(J)=M(I)
2200      REM Mutate
2300      IF C(J)=1 THEN   T=3
2400      IF C(J)=2 THEN   T=4
2500      IF C(J)=3 THEN   T=1
2600      IF C(J)=4 THEN   T=2
2700      C(J)=T
2800      J=J+1
2900    NEXT I
3000    REM COP Enzyme reads last generation
3100    FOR I=1 TO S1
3200      C(J)=L(I)
3300      J=J+1
3400    NEXT I
3500    REM Print current line
3600    FOR I=1 TO J
3700      IF C(I)=1 THEN PRINT "G"
3800      IF C(I)=2 THEN PRINT "C"
3900      IF C(I)=3 THEN PRINT "A"
4000      IF C(I)=4 THEN PRINT "T"
4100    NEXT I
4150    J=J-1
4200    PRINT " Length: ", J
```

```
4300    IF J>P THEN P=J
4400    REM CLEAV splits sequence at T-T
4500    S3=J
4600    FOR I=1 TO S3
4700      IF C(I)=4 THEN 4750
4720      GOTO 4800
4750      IF C(I+1)=4 THEN  GOSUB 9000
4800    NEXT I
4900    REM CONTINUED....
5000    REM BASIC Cancer Game Program (CONTINUED)
6800    REM Update by swapping generations
6900    FOR I=1 TO S1
7000      M(I)=L(I)
7100    NEXT I
7200    S2=S1
7300    FOR I=1 TO S3
7400      L(I)=C(I)
7500      S1=S3
7550    NEXT I
7600 NEXT G
7700 PRINT "Maximum string length found is "; P
7800 STOP
9000 REM New length because of cleavage
9100 S3=I
9200 PRINT "Cleavage has occurred"
9300 FOR I=1 TO S3
9400    IF C(I)=1 THEN PRINT "G"
9500    IF C(I)=2 THEN PRINT "C"
9600    IF C(I)=3 THEN PRINT "A"
9700    IF C(I)=4 THEN PRINT "T"
9800 NEXT I
9900 PRINT
9990 RETURN
9999 END
```

Chapter 17 Infinite Star Chambers

C Code

```c
/* Compute Coordinates of Stars within Stars */
#include <math.h>
#include <stdio.h>
main()
{
  float r,pi,x[10],y[10],theta[10];
  int i,j;
  pi=3.1415; r=1;
  theta[1]=pi/2;
  /* Print 3 nestings of stars */
  for (j=1; j<=3; j++) {
    /* Compute coordinates of stars */
    for (i=1; i<=7; i++) {
        x[i]=r*cos(theta[i]);
        y[i]=r*sin(theta[i]);
```

```
            theta{i+1}=theta{i}+(72*pi/180.0);
      }
      for (i=1; i<=5; i++) {
            /* Print coordinates of each line segment */
            printf("%f %f\n", x[i],y[i]);
            printf("%f %f\n", x[i+2],y[i+2]);
      }
      r=r/2.618;
      theta[1]=theta[1]+(36*pi/180.0);
   }
}
```

BASIC Code

```
10 REM Compute Coordinates of Stars within Stars
20 DIM X(10), Y(10), T(10)
30 P=3.1415
40 R=1
60 T(1)=P/2
70 REM Print 3 nestings of stars
80 FOR J=1 TO 3
90   REM Compute coordinates of stars
100  FOR I=1 TO 7
110      X(I)=R*COS(T(I))
120      Y(I)=R*SIN(T(I))
130      T(I+1)=T(I)+(72*P/180.0)
140  NEXT I
150  FOR I=1 TO 5
160      REM Print coordinates of each line segment
180      PRINT X(I),Y(I)
190      PRINT X(I+2),Y(I+2)
200  NEXT I
210  R=R/2.618
220  T(1)=T(1)+(36*P/180.0)
230  NEXT J
240 END
```

C Code

```
/* Compute Length of Stars */
#include <math.h>
#include <stdio.h>

main()
{
float Length, r;
int i;

   r=(3.0 + sqrt(5.0))/2.0;
   /* Compute length of star's edge */
   for (i=0; i<=65; i++) {
         Length=pow(r,i);
         printf("%d %g %g\n",i,Length,Length/12.);
   }
   /* Compute cumulative length of all star edges */
```

```
printf("**************\n");
for (i=0; i<=65; i++) {
     Length=5*(pow(r,i)-1)/(r-1);
     printf("%d %g %g\n",i,Length,Length/12.);
}
}
```

BASIC Code

```
10 REM Compute Star Lengths
20 R=(3.0+SQR(5.0))/2.0
30 REM Compute length of star's edge
50 FOR I=0 TO 65
60    L=R**I
70    PRINT I, L
80 NEXT I
90 REM Compute cumulative length of all star edges
100 PRINT "**************"
110 FOR I=0 TO 65
120    L=5*((R**I)-1)/(R-1)
130    PRINT I, L
140 NEXT I
150 END
```

Chapter 18 Infinitely Exploding Circles

C Code

```
/* Exploding Circles */
#include <math.h>
#include <stdio.h>
#define SQR(x) ((x)*(x))
#define QUAD(x) ((x)*(x)*(x)*(x))

int n;
double pi=M_PI, prod, pi2, pi4, A;

main()
{
    /* Compute limiting radius using simple method 1 */
    prod=1;
    for (n=3; n<100; n++) {
       prod=prod*(1.0/cos(pi/n));
       /* print out radius values as circles expand */
       if((n%10)==0) printf("%d %f\n", n-2,prod);
    }
    printf("Final Result, Method 1, is: %d %f\n", n-2,prod);

    /* Compute limiting radius using method 2 */
    pi2=pi*pi;
    pi4=pi2*pi2;

    A=((1-pi2/2+pi4/24)*(1-pi2/8+pi4/384)*pi4/sqrt(24))/
       (sin(pi2/sqrt(6+2*sqrt(3)))*sin(pi2/sqrt(6-2*sqrt(3))));
       printf("A is: %f\n",A);
```

```
      prod=1;
      for (n=3; n<100; n++) {
         prod=prod*((1-pi2/(2*SQR(n)) +
         pi4/(24*QUAD(n)))/cos(pi/n));
      }
      prod=A*prod;
      printf("Final Result, Method 2, is: %d %f\n",n-2,prod);
}
```

BASIC Code

```
10  REM Exploding Circles
20  P=3.1415926535
30  REM Compute limiting radius using simple method 1
40  Q=1.0
50  FOR N=3 TO 100
60    Q=Q*(1.0/COS(P/N))
70    REM Print out radius values as circles expand
80    PRINT N-2,Q
85  NEXT N
90  PRINT "Final Result, Method 1, is:",N-2,Q
100 REM Compute limiting radius using method 2
110 P2=P*P
120 P4=P2*P2
130 X=((1-P2/2+P4/24)*(1-P2/8+P4/384)*P4/SQR(24))
140 Y=(SIN(P2/SQR(6+2*SQR(3)))*SIN(P2/SQR(6-2*SQR(3))))
150 A=X/Y
160 PRINT "A is:",A
170 Q=1.0
180 FOR N=3 TO 100
190   Q=Q*((1-P2/(2*N*N)+P4/(24*N**4))/COS(P/N))
200 NEXT N
210 Q=A*Q
220 PRINT "Final Result, Method 2, is:", N-2,Q
230 END
```

Chapter 19 The Infinity Worms of Callisto

C Code

The following code allows you to compute the first ten values for the various infinity-worm series.

```
/* Compute Infinity Worm Series */
#include <math.h>
#include <stdio.h>

main()
{

    float sum1, sum2, sum3, sum4, sum5, sum6;
    int i;
    sum1=0; sum2=0; sum3=0;
    sum4=0; sum5=0; sum6=0;
```

```c
    /* Compute first ten terms */
    for (i=1; i<=10; i++) {
        /* Green worm */
        sum1=sum1+1/(log(log(i+2)));
        printf("Green Worm: %d %f\n",i,sum1);
        /* Blue worm */
        sum2=sum2+1/(log(i+1));
        printf("Blue Worm: %d %f\n",i,sum2);
        /* Violet worm */
        sum3=sum3+1/sqrt(i);
        printf("Violet Worm: %d %f\n",i,sum3);
        /* Crimson worm */
        sum4=sum4+1/(float)i;
        printf("Crimson Worm: %d %f\n",i,sum4);
        /* Black worm */
        sum5=sum5+1/((i+1)*log(i+1));
        printf("Black Worm: %d %f\n",i,sum5);
        /* Transparent worm*/
        sum6=sum6+1/((i+2)*log(i+2)*log(log(i+2)));
        printf("Transparent Worm: %d %f\n",i,sum6);
    }
} /* end program */
```

BASIC Code

```basic
10 REM Compute first Ten Terms of Worm Series
20 S1=0
30 S2=0
40 S3=0
50 S4=0
60 S5=0
70 S6=0
80 REM Compute first 10 terms
90 FOR I=1 TO 10
100    REM Green worm
110    S1=S1+1/(LOG(LOG(I+2)))
120    PRINT "Green Worm: ",I,S1
130    REM Blue worm
140    S2=S2+1/(LOG(I+1))
150    PRINT "Blue Worm: ",I,S2
160    REM Violet worm
170    S3=S3+1/SQR(I)
180    PRINT "Violet Worm: ",I,S3
190    REM Crimson worm
200    S4=S4+1/I
210    PRINT "Crimson Worm: ",I,S4
220    REM Black worm
230    S5=S5+1/((I+1)*LOG(I+1))
240    PRINT "Black Worm: ",I,S5
250    REM Transparent worm
260    S6=S6+1/((I+2)*LOG(I+2)*LOG(LOG(I+2)))
270    PRINT "Transparent Worm: ",I,S6
280 NEXT I
290 END
```

Chapter 21 The Fractal Golden Curlicue Is Cool

C Code

```
/* Generate Curlicues */
#include <math.h>
#define SGN(x) ((x)<0 ? -1 : (x)>0 ? 1 : 0)
float x[2000], y[2000];
float pi, pi2, cons, ang1, step, ang2, ang3;
int i;

main()
{
 pi=3.14159; pi2=2.0*pi;
 /* Golden section, as an example constant */
 cons=(sqrt(5.0)+1.0)/2.0;
 ang1=pi2*cons; /* primary angle */
 step=1;        /* step size */
 ang3=0; ang2=0;
 x[0]=y[0]=0;
 for (i=0; i<=1000; i++) {
     ang2=ang2+ang1;
     /* Prevent angle from growing huge: */
     while(ang2>pi2) ang2=ang2-pi2;
     ang3=ang3+ang2;
     while(fabs(ang3)>pi2)
     ang3=ang3-SGN(ang3)*pi2;
     x[i+1]=x[i]+step*cos(ang3);
     y[i+1]=y[i]+step*sin(ang3);
     /* Write out coordinates of line segments */
     printf("%f %f\n",x[i+1],y[i+1]);
 } /* i */
}
```

BASIC Code

```
10 REM Compute Coordinates of Curlicues
20 P1=3.14159
30 P2=2.0*P1
40 REM Golden section, as an example constant
50 C=(SQR(5.0)+1.0)/2.0
55 REM Primary angle
60 A1=P2*C
70 REM Step size
80 S=1
90 A3=0
100 A2=0
105 X=0
107 Y=0
110 FOR I = 0 TO 1000
120     A2=A2+A1
130     REM Prevent angle from growing huge:
140     FOR J=1 TO 100
150          IF A2<=P2 GOTO 180
```

```
160          A2=A2-P2
170      NEXT J
180      A3=A3+A2
190      FOR J=1  TO 100
200          IF ABS(A3)<=P2 GOTO 230
210          A3=A3-SGN(A3)*P2
220      NEXT J
230      X=X+S*COS(A3)
240      Y=Y+S*SIN(A3)
250      REM Write out coordinates of line segments
260      PRINT X, Y
270 NEXT I
280 END
```

Chapter 22 The Loneliness of the Factorions

C Code

Youry Vymenets of St. Petersburg University, Russia, says, "I'm afraid your factorions will be lonely forever. This is sad, but a lot of people rest alone throughout
their entire life without even a single companion. . . ." He goes on to prove this statement (about factorions, not about life), and has given me permission to reproduce his
C code for brute force search of factorions. The C program searches all numbers less
than 10^8.

```c
/* Loneliness of the Factorions (vymenets) */
#include <stdio.h>
#include <math.h>
 long i,j,n,m,s;
 long a[9+1][1+1];
 long b[7+1];
 long c[7+1];
main()

{
  m=10000000;
  a[0][1]=1;
  a[1][1]=1;
  a[2][1]=2;
  a[3][1]=6;
  a[4][1]=24;
  a[5][1]=120;
  a[6][1]=720;
  a[7][1]=5040;
  a[8][1]=40320;
  a[9][1]=362880;
  for (i=0; i<=7; i++)
    {
      b[i]=0;
      c[i]=0;
    }
  for (i=0; i<=9; i++)
    a[i][0]=0;
```

```
    for (j=1; j<=m; j++)
      {b[0]++; c[0]=1;
       if(b[0]==10) {b[0]=0; b[1]++; c[1]=1;
       if(b[1]==10) {b[1]=0; b[2]++; c[2]=1;
       if(b[2]==10) {b[2]=0; b[3]++; c[3]=1;
       if(b[3]==10) {b[3]=0; b[4]++; c[4]=1;
       if(b[4]==10) {b[4]=0; b[5]++; c[5]=1;
       if(b[5]==10) {b[5]=0; b[6]++; c[6]=1;
       if(b[6]==10) {b[6]=0; b[7]++; c[7]=1;
                     }}}}}}}
      s=0;
      for (i=0; i<=7; i++)
        s=s+a[b[i]][c[i]];
      if(s==j) printf("%ld,",j);
    }
}
```

The program works for several minutes and yields only the two known multidigit factorions.

Herve Bronnimann from Princeton University has also conducted research in the existence of factorions. The following is his C code for searching for such numbers.

```
/* Loneliness of the Factorions */
/* Herve Bronnimann            */
#include <stdio.h>
#define Base  10
main()
{
  long Functorial[Base], MaxFunctorial;
  long MaxDigits, PowBase;
  long j, k, l;

  /* Modify these for other functorions */
  Functorial[0]=1;
  for (j=1; j<Base; j++)
    Functorial[j]=j*Functorial[j-1];
  MaxFunctorial=Functorial[Base-1];
  /* Compute the maximum number of digits */
  MaxDigits=1; PowBase=Base;
  while(MaxFunctorial*MaxDigits>PowBase) {
    MaxDigits++;
    PowBase=PowBase*Base;
  }
  /* Compute the factorions */
  printf("MaxDigits = %ld\n",MaxDigits);
  printf("PowBase   = %ld\n",PowBase);
  printf("MaxFunct  = %ld\n",MaxFunctorial);
  for (j=0; j<PowBase; j++) {
    for (k=j, l=0; k>0; k/=Base)
      l += Functorial[k%Base];
    if(j==l)
      printf("Found a factorion (written in base 10): %ld\n",j);
  }
}
```

BASIC Code

The following allows you to search for narcissistic numbers, as described in the text.

```
10 REM Search for Cubical Narcissistic Numbers
15 REM Search all 3-digit integers
20 FOR I=100 TO 999
30    A=INT(I/100)
40    B=INT(I/10)-10*A
50    C=I-100*A-10*B
55    IF I<>A**3+B**3+C**3 THEN 90
60    PRINT "Narcissistic Number";I
70    PRINT "Equals";A**3;"+";B**3;"+";C**3
80    PRINT
90 NEXT I
100 END
```

Chapter 23 Escape from Fractalia

BASIC Code

This program, ESCAPE.BAS, automatically produces an unlimited number of escape-time fractals similar to those shown in Chapter 23.

```
REM Program ESCAPE.BAS by J. C. Sprott
DIM a(12)                        'Array of coefficients
RANDOMIZE TIMER                  'Reseed random numbers
SCREEN 12                        'Assume VGA graphics
n% = 0
WHILE INKEY$ = ""                'Loop until a key is pressed
    IF n% = 0 THEN CALL setparams(x, y)
    CALL advancexy(x, y, n%)
    CALL testsoln(x, y, n%)
    IF n% = 1000 THEN CALL display: n% = 0
WEND
END

SUB advancexy (x, y, n%)         'Advance (x, y) at step n%
SHARED a()
xnew = a(1) + x * (a(2) + a(3) * x + a(4) * y) + y * (a(5) + a(6)
    * y)
y = a(7) + x * (a(8) + a(9) * x + a(10) * y) + y * (a(11) + a(12)
    * y)
x = xnew
n% = n% + 1
END SUB

SUB display ()                   'Plot escape-time contours
FOR i% = 0 TO 639
    FOR j% = 0 TO 479
        x = -5 + i% / 64
        y = 5 - j% / 48
        n% = 0
        WHILE n% < 128 AND x * x + y * y < 1000000
```

```
          CALL advancexy(x, y, n%)
     WEND
     PSET (i%, j%), n% MOD 16
  NEXT j%
NEXT i%
END SUB

SUB setparams (x, y)              'Set a() and initialize (x,y)
SHARED a()
x = 0: y = 0
FOR i% = 1 TO 12: a(i%) = (INT(25 * RND) - 12) / 10: NEXT i%
END SUB

SUB testsoln (x, y, n%)          'Test the solution
IF n% = 1000 THEN n% = 0          'Solution is bounded
IF x * x + y * y > 1000000 THEN   'Solution escaped
     IF n% > 100 THEN n% = 1000 ELSE n% = 0
END IF
END SUB
```

Chapter 24 Are Infinite Carotid-Kundalini Functions Fractal?

C Code

```c
/* Compute Carotid-Kundalini Curves */
#include <math.h>
#include <stdio.h>

main()
{
  float x,y;
  int n;

  /* Superimpose 25 curves */
  for (n=1; n<=25; n=n+1) {
      for (x=-1; x<=1; x=x+.01) {
          y=cos((float)n*x*acos(x));
          /* Write out x, y points for plotting */
          printf("%f %f n",x,y);
      }
  }
}
```

BASIC Code

```
10 REM Compute Carotid-Kundalini Curves
20 REM Superimpose 25 curves
30    FOR N=1 TO 25
40        FOR X=-1 TO 1 STEP 0.01
50            Y=COS(N*X*ACOS(X))
60            REM Write out x, y points for plotting
70            PRINT X, Y
```

```
80        NEXT X
90      NEXT N
100 END
```

BASIC Code for ACOS

As noted earlier in this chapter, some versions of BASIC do not have an ACOS function. Included here is a program by William Rudge that computes this function. It should work in QBASIC or QuickBASIC.

```
' Main program to test the ACOS# function
' On my computer with QBASIC, it's accurate to six digits.
DECLARE FUNCTION ACOS# (y AS DOUBLE)
COMMON HALF.PI#, PI#
HALF.PI = 2# * ATN(1)
PI = 4# * ATN(1)
PRINT
PRINT
PRINT "pi/2 = "; HALF.PI, "pi = "; PI
INPUT a$
i# = 0
WHILE i# <= PI
  v# = ACOS#(COS(i#))
  '             actual      computed     difference
  PRINT USING "###.###### ###.###### +#.########"; i#; v#; i# - v#
  i# = i# + .01
WEND
END


' Note that y ^ 2 is same as y * y.
FUNCTION ACOS# (y AS DOUBLE) STATIC
SHARED HALF.PI, PI
  IF ABS(y) < .5 THEN
    v# = HALF.PI - ATN(y / SQR(1 - y ^ 2))
  ELSE
    v# = ATN(SQR(1 - y ^ 2) / y)
    IF y < 0 THEN v# = PI + v#
  END IF
  ACOS# = v#
END FUNCTION
```

Chapter 25 The Crying of Fractal Batrachion 1,489

C Code

```
/* Compute Batrachions */
#include <math.h>
#include <stdio.h>

main()
{
  int curvename, renorm, i, n, a[20000], look[21];
```

```
float x, ratio, s;
curvename=1; renorm=0; x=0
/* Generate lookup table for renormalization */
if(renorm==1)
for (i=1; i<20; i++) look[i]=pow(2,i);
/* Initial seeds */
a[1]=1; a[2]=1;
for (n=3; n<2048; n++) {
  /* Conway's batrachion */
  if(curvename==1)
    a[n]=a[a[n-1]]+a[n-a[n-1]];
  /* Hofstadter's */
  if(curvename==2)
    a[n]=a[n-a[n-1]]+a[n-a[n-2]];
  /* Mallows' */
  if(curvename==3)
    a[n]=a[a[n-2]]+a[n-a[n-2]];
  /* Compute renormalization scale factor for plot */
  if(renorm==1) {
    x++;
    for (i=1; i<20; i++)
        if(n==look[i]) {x=0; s=1./pow(2,i);}
  }
  ratio=(float)a[n]/(float)n;
  /* Generate data points for plot */
  if(renorm==0)    printf("%d %f\n",n,ratio);
  /* Generate data points for renormalization plot */
  if(renorm==1) printf("%f %f\n",x*s,ratio);
} /* for */
} /* end program */
```

BASIC Code

```
10 REM Compute Batrachions
20 DIM A(3000), L(21)
30 C=1
40 R=0
45 X=0
50 REM Generate lookup table for renormalization
60 IF R=0 THEN GOTO 100
70 FOR I=1 TO  20
80     L(I)=2**I
90 NEXT I
100 REM Initial seeds
110 A(1)=1
120 A(2)=1
130 FOR N=3 TO  2048
140    REM  Conway's batrachion
150    IF C=1 THEN A(N)=A(A(N-1))+A(N-A(N-1))
160    REM Hofstadter's
170    IF C=2 THEN A(N)=A(N-A(N-1))+A(N-A(N-2))
180    REM Mallows'
190    IF C=3 THEN A(N)=A(A(N-2))+A(N-A(N-2))
```

```
200    REM Compute renormalization scale factor for plot
210    IF R=0 THEN GOTO 300
220    X=X+1
230    FOR I=1 TO 20
240        IF N=L(I) THEN X=0
250        S=1/(2**I)
260    NEXT I
300    R1=A(N)/N
310    REM Generate data points for plot
320    IF R=0 THEN PRINT N, R1
330    REM Generate data points for renormalization plot
340    IF R=1 THEN PRINT X*S, R1
350 NEXT N
360 END
```

Chapter 27 Recursive Worlds

C Code

The C code for this chapter allowed me to compute the structure for the coral-like form. To achieve the bumpy effect, I used a separate graphics program to draw spheres at coordinates determined by the algorithm implemented in C.

```c
/* Compute Coral Form */
#include <device.h>
#include <math.h>
#include <stdio.h>
#define pi 3.1415927
int ang_control,seed;
float depth,z,red,green,blue,radius;
main()
{
    int n;
    srand(12345); /* set random number seed */
    /* Angle controls branch pattern */
    printf("Enter angle in degrees (0<angle<45) n");
    scanf("%d", &ang_control);
    printf("Give recursion depth, n n");
    scanf("%d",&n); depth=(float)n;
    branch(n,0.,0.,1.,0.);
}
branch(n,xinit,yinit,a,theta)
int n; float xinit,yinit,a,theta;
{
    float x[5],y[5],xx[5],yy[5],ctheta,stheta,c1,c2,b,c,
    float angle,cangle,sangle;
    int i,fuzziness;       /* theta and angle in radians */
    if(n==0) return;       /* ang_control in degrees    */
    /* Use random numbers to control natural growth      */
    fuzziness=rand()%(2*ang_control+1)-ang_control;
    angle=(45+fuzziness)*pi/180.;
    x[0]=x[3]=xinit; x[1]=x[2]=xinit+a;
```

```
y[0]=y[1]=yinit; y[2]=y[3]=yinit+a;
cangle=cos(angle); sangle=sin(angle);
c=a*angle; b=a*sangle;
x[4]=x[3]+c*cangle; y[4]=y[3]+c*sangle;
/* Rotate about xinit, yinit through angle theta */
ctheta=cos(theta); stheta=sin(theta);
c1=xinit-xinit*ctheta+yinit*stheta;
c2=yinit-xinit*stheta-yinit*ctheta;
/* Compute color for each ball */
red=1-(float)n/depth; blue=(float)n/depth; green=1;
z=0;
for (i=0; i<5; i++) {
    xx[i]=x[i]*ctheta-y[i]*stheta+c1;
    yy[i]=x[i]*stheta+y[i]*ctheta+c2;
} /* endfor */
for (i=0; i<5; i++) {
    /* Compute radius for each ball */
    radius=fabs((xx[3]-xx[2]))/1.2;
    /* For each ball, dump radius and color */
    printf("%f %f %f %f %f %f %f\n",
           xx[i],yy[i],z,radius,red,green,blue);
}
/* Now you must find a graphics program that will
   use this data to draw spheres or circles. */

/* Recursive calls */
branch(n-1,xx[3],yy[3],c,theta+angle);
branch(n-1,xx[4],yy[4],b,theta+angle-0.5*pi);
}
```

Chapter 29 Cyclotron Puzzles

C Code

This short piece of code generates the puzzle.

```
/* Generate Cyclotron Puzzle */
#include <math.h>
#include <stdio.h>

int x[100][100], i, j, k;

main()
{
  srand(123456789); /* random number gen seed */
  /* generate array of numbers */
  k=0;
  for (i=0; i<15; i++) {
    for (j=0; j<20; j++) {
      k++;
      /* generate random numbers (0 - 100) */
      x[i][j]=((float)rand()/32767.)*100.;
      /* create hole in middle */
```

```
    if((j>2) && (j<17) && (i>2) && (i<12)) printf("    ");
    else   printf("%2d ",x[i][j]);
    if(((k)%20)==0) printf(" n");
    } /* for */
  } /* for */
}
```

For Further Exploration

Included in this appendix are listings of longer computer programs for various topics discussed in this book.

Recursive C Program to Explore Worm World

```
/*===============================================================
**                WORM WORLD EXPLORER PROGRAM
**                     by Stephen Tavener
**
** usage: worm infile outfile x y
**
** infile  is a list of HEIGHT rows of WIDTH numbers (default: stdin)
** outfile is the results file (default: stdio)
** x,y     are the start coordinates if you don't want to check
**         every position
**===============================================================*/
#include <stdio.h>
#include <string.h>
#define WIDTH     8
#define HEIGHT    8
#define MAX_VALS 25
long SeqLen = 0;
char StoredPath[MAX_VALS];
char Path[MAX_VALS];
FILE *InFile;
FILE *OutFile;
int SquareValues[HEIGHT][WIDTH];
char ValuesGone[MAX_VALS] = {0,0,0,0,0,
                             0,0,0,0,0,
                             0,0,0,0,0,
                             0,0,0,0,0,
                             0,0,0,0,0};
void Recurse (long Depth, long X, long Y)
{
  /* Can we go here ? */
  if (ValuesGone[SquareValues[Y][X]]) return;
  /* Set up... */
  ValuesGone[SquareValues[Y][X]] = 1;
  Depth++;
```

```
  if (Depth > SeqLen)
  {
     SeqLen = Depth;
     memcpy (StoredPath, Path, sizeof (Path));
  }
  /* Recurse... */
  if (X-1 >= 0)       { Path[Depth-1] = 'L'; Recurse (Depth, X-1, Y); }
  if (X+1 < WIDTH)    { Path[Depth-1] = 'R'; Recurse (Depth, X+1, Y); }
  if (Y-1 >= 0)       { Path[Depth-1] = 'U'; Recurse (Depth, X, Y-1); }
  if (Y+1 < HEIGHT)   { Path[Depth-1] = 'D'; Recurse (Depth, X, Y+1); }
  /* Clean up... */
  ValuesGone[SquareValues[Y][X]] = 0;
}
/* Continued */
/* ...CONTINUED */ int main (int argc, char *argv[])
{
  long   StartX = -1;
  long   StartY = -1;
  long   HtLoop, WdLoop;
  char   LineBuf[255];
  InFile  = stdin;
  OutFile = stdout;
  if (argc > 4) sscanf (argv[4], "%d", &StartY);
  if (argc > 3) sscanf (argv[3], "%d", &StartX);
  if (argc > 2) OutFile = fopen (argv[2], "w");
  if (argc > 1) InFile  = fopen (argv[1], "r");
  for (HtLoop = 0; HtLoop < HEIGHT; HtLoop++)
     for (WdLoop = 0; WdLoop < WIDTH; WdLoop++)
        fscanf (InFile, "%d", &SquareValues[HtLoop][WdLoop]);

  if (StartY < 0 || StartX < 0)
  {
     for (StartY = 0; StartY < HEIGHT; StartY++)
     {
        for (StartX = 0; StartX < WIDTH; StartX++)
        {
           SeqLen = 0;
           Recurse (0, StartX, StartY);
           fprintf (OutFile,
              "Longest Sequence from (%0d, %0d) is %*.*s
                 (length: %0d) n",
                    StartX, StartY, SeqLen-1, SeqLen-1, StoredPath,
                    SeqLen);
        }
     }
  }
  else
  {
     Recurse (0, StartX, StartY);
     fprintf (OutFile,
        "Longest Sequence from (%0d, %0d) is %*.*s (length:
           %0d) n",
           StartX, StartY, SeqLen-1, SeqLen-1, StoredPath,
           SeqLen);
```

```
    }

    return 0;
}
```

Gofer Program to Search for Vampire Numbers

Tony Davie of St. Andrews University's Department of Mathematical and Computational Sciences used Gofer to conduct research on the vampire numbers. He mentioned to me that Gofer is a functional programming language, not to be confused with Gopher, an international network hyperprogramming navigation system. Gofer is very powerful because of its so-called "laziness," which is the feature that allows the use of infinite objects.[1]

Let us look at Tony's Gofer program for finding vampires:

```
vamps = [(x,y,z) | x <- [1..[, sx=show x, y <- [1..x[,
                   z = x*y, sort(show z)== sort(sx++show y)[
```

You can interpret this program as follows: "vamps" is the (infinite) list of triples (x, y, z) such that x is drawn from ("<-") the infinite list of positive integers "[1..[." The term sx is the string (list) of characters representing x in decimal "show x." The term y is drawn from ("<-") the list of integers from 1 up to x. The term z is the product of x and y. Finally, there is a condition that filters all these possibilities, retaining only the ones where the sorted digits of z, the product (sort(show z)), are the same as (= =) the sorted digits of x appended to (++) those of y (sort(sx++show y)). The "= =" is used for testing equality whereas the "=" is used for definitional equality.

Tony says readers may think of this in much the same way as a group of "for" loops in a more conventional computer language, though the idea of infinite lists probably has no parallel:

```
            for x=1 to infinity do
            begin
                    let sx=show(x);
                    for y=1 to x do
                    begin
                            let z=x*y
                            if sort(show(z))==sort(sx++show(y)) then
                                    (x,y,z) is a candidate
                    end
            end
```

Tony gives another example for Pythagorean triples:

```
pyth = [(x,y,z) | x <- [1..[, y <- [1..x[, z <- [1..y[, x*x
       == y*y + z*z[
```

which is a good specification as well as a working program, but a more efficient one would be:

```
pyth' = [(m*m+n*n, 2*m*n, m*m-n*n) | m <- [2..[, n <- [1..m-1[ [
```

and readers might like to add a filtering condition so there is no common factor in the triple.

Here is another example from Tony showing off the "laziness" feature of Gofer that allows the use of infinite objects. This computes an infinite list of powers of 2:

```
powers 1
  where powers n = n : powers (2*n)
```

This can be "expanded" as follows:

```
powers 1 = 1 : powers 2
         = 1 : 2 : powers 4
         = 1 : 2 : 4 : powers 8
         = ...
```

The ":" operator joins a single new element onto a list of elements of the same type. It is right-associative. Tony notes that there is a very good introduction to laziness and its uses in J. Hughes, "Why Functional Programming Matters," *Computer Journal*, 32(2) (1989).

As additional research in vampire numbers, Tony left (a slightly modified) version of his Gofer program running for a long time on his computer:

```
vamps = [(x,y,z) | x <- [1..[, sx=show x, y <- [1..x[,
                   z = x*y, sort(show z)== sort(sx++show y)[


lineout (x:xs) = show3 x ++ ' n':lineout xs

show3 (x,y,z) = show x ++ '*':show y ++ '=': show z
```

Using this program, the largest true vampire he found was:

$$986 \times 953 = 939{,}658$$

The largest pseudovampire he found was:

$$2{,}573 \times 986 = 2{,}536{,}978$$

REXX Program to Search for Vampire Numbers

Steve Frye from North Carolina used the following REXX program on a VM system to search for vampire numbers.

```
/* Searching for Vampire Numbers */

outfile="vampire numbers al"        /* output file name */
ProgressInterval=10                 /* How often to print a
                                       progress indicator */

say "Search for Vampires!"
say "Starting at: "
pull starting
```

```
say "Ending at: "
pull ending

if starting='' then starting=1000
if ending='' then ending=9999
if ending<1000 then
  do
    say "There are no vampires under 1000"
    goto EndOfProgram
  end

MinVampireDigits=length(starting)
if MinVampireDigits//2 = 1 then MinVampireDigits=MinVampireDigits+1
MaxVampireDigits=length(ending)
if MaxVampireDigits//2=1 then MaxVampiredigits=MaxVampireDigits-1
line="Searching for vampires greater than "
line=line||starting||" and less than "||ending
'execio 1 diskw ' outfile '( VAR LINE )'
VampireCount=0

numeric digits MaxVampireDigits+2

do VampireDigits=MinVampireDigits to MaxVampiredigits by 2
  TargetRangeMin=10**(VampireDigits-1)
  if TargetRangeMin < starting then TargetRangeMin=starting
  TargetRangeMax=10*TargetRangeMin - 1
  if TargetRangeMax > ending then TargetRangeMax=ending
  BiggestFactor=(10**(VampireDigits/2))-1
  InitialFirstFactor=TargetRangeMin%BiggestFactor
  line=" "
  'execio 1 diskw ' outfile '( VAR LINE )'
  line="Looking in the range" TargetRangeMin "to" TargetRangeMax
  'execio 1 diskw ' outfile '( VAR LINE )'
  line=" "
  'execio 1 diskw ' outfile '( VAR LINE )'
  say "Looking in the range" TargetRangeMin "to" TargetRangeMax
  say "First Factor will range from" InitialFirstFactor "to" Biggest-
Factor
  do i=InitialFirstFactor to BiggestFactor
    if i//ProgressInterval=0 then say i
    do j=i to BiggestFactor
      product=i*j
      if product < TargetRangeMin then iterate
      if product > TargetRangeMax then leave
      call VampireCheck
      if vampire="no" then iterate
      call DuplicateCheck
      line=i||" x "||j||" = "||product
      if duplicate="yes" then line=line||" duplicate"
      'execio 1 diskw ' outfile '( VAR LINE )'
      line=insert(" "," ",1,VampireDigits/2+2," ")||line
      say line
    end
```

```
        end
      end
   say " "
   say "There were" VampireCount "distinct vampire numbers found."

   EndOfProgram:
   finis outfile
   exit
/****************************************************************/
   VampireCheck:
      testnum=product
      vampire="yes"
      do FactorDigit=1 to length(i)
         ptr=pos(substr(i,FactorDigit,1),testnum)
         if ptr=0 then
            do
               vampire='no'
               leave
            end
         else
            do
               testnum=overlay(" ",testnum,ptr)
            end
         ptr=pos(substr(j,FactorDigit,1),testnum)
         if ptr=0 then
            do
               vampire='no'
               leave
            end
         else
            do
               testnum=overlay(" ",testnum,ptr)
            end
      end

   return

/****************************************************************/
   DuplicateCheck:
      duplicate="no"
      if vampires.product=1 then
         do
            duplicate="yes"
         end
      else
         do
            VampireCount=VampireCount+1
            vampires.product=1
         end

   return
```

REXX Program to Search for Vampire Numbers (OS/2)

Steve Frye from North Carolina also used the following REXX program on an OS/2 system to search for vampire numbers.

```
/* Searching for Vampire Numbers */

outfile="d: vampire.num"              /* output file name */
ProgressInterval=10                   /* How often to print a
                                         progress indicator */

say "Search for Vampires!"
say "Starting at: "
pull starting
say "Ending at: "
pull ending

if starting='' then starting=1000
if ending='' then ending=9999
if ending<1000 then
   do
      say "There are no vampires under 1000"
      goto EndOfProgram
   end

MinVampireDigits=length(starting)
if MinVampireDigits//2 = 1 then MinVampireDigits=MinVampireDigits+1
MaxVampireDigits=length(ending)
if MaxVampireDigits//2=1 then MaxVampiredigits=MaxVampireDigits-1
call lineout outfile,"Searching for vampires greater than" starting
"and less than" ending
VampireCount=0

numeric digits MaxVampireDigits+2

do VampireDigits=MinVampireDigits to MaxVampiredigits by 2
   TargetRangeMin=10**(VampireDigits-1)
   if TargetRangeMin < starting then TargetRangeMin=starting
   TargetRangeMax=10*TargetRangeMin - 1
   if TargetRangeMax > ending then TargetRangeMax=ending
   BiggestFactor=(10**(VampireDigits/2))-1
   InitialFirstFactor=TargetRangeMin%BiggestFactor
   call lineout outfile," "
   call lineout outfile,"Looking in the range" TargetRangeMin "to"
     TargetRangeMax
   call lineout outfile," "
   say "Looking in the range" TargetRangeMin "to" TargetRangeMax
   say "First Factor will range from" InitialFirstFactor "to"
     BiggestFactor
   do i=InitialFirstFactor to BiggestFactor
      if i//ProgressInterval=0 then say i
      do j=i to BiggestFactor
        product=i*j
        if product < TargetRangeMin then iterate
        if product > TargetRangeMax then leave
```

```
      call VampireCheck
      if vampire="no" then iterate
      call DuplicateCheck
      line=i||" x "||j||" = "||product
      if duplicate="yes" then line=line||"  duplicate"
      call lineout outfile,line
      line=insert(" "," ",1,VampireDigits/2+2," ")||line
      say line
    end
  end
end
say " "
say "There were" VampireCount "distinct vampire numbers found."

EndOfProgram:
exit
/*****************************************************************/
VampireCheck:
  testnum=product
  vampire="yes"
  do FactorDigit=1 to length(i)
    ptr=pos(substr(i,FactorDigit,1),testnum)
    if ptr=0 then
        do
           vampire='no'
           leave
        end
    else
        do
           testnum=overlay(" ",testnum,ptr)
        end
    ptr=pos(substr(j,FactorDigit,1),testnum)
    if ptr=0 then
        do
           vampire='no'
           leave
        end
    else
        do
           testnum=overlay(" ",testnum,ptr)
        end
  end
return

/*****************************************************************/
DuplicateCheck:
  duplicate="no"
  if vampires.product=1 then
    do
       duplicate="yes"
       leave
    end
  else
    do
       VampireCount=VampireCount+1
```

```
        vampires.product=1
      end

return
```

Another REXX Program to Search for Vampire Numbers

Glenn S. Knickerbocker from New York used the following REXX program running on IBM VM to search for vampire numbers.

```
/* Glenn S. Knickerbocker, GSKNICK at FSHVMFK1, 2 May 1994 */
/* Find Cliff's "vampire" numbers                          */
Arg start max min '(' exact
If min = '' then min = start
start = start - start//9
max = max + max//9
min = min - min//9
exact = Abbrev('EXACT', exact, 1)

Do i = start to max by 9
  Do j = min to i by 9
    Call vampire i   , j
    Call vampire i+2, j+2
    Call vampire i+3, j+6
    Call vampire i+5, j+8
    Call vampire i+6, j+3
    Call vampire i+8, j+5
  End
End

Exit

Vampire:  Procedure Expose exact
  Parse Arg x, y
  If exact then If length(x) ¬= length(y) then Return
  c = x || y
  p = x * y
  If length(c) ¬= length(p) then Return
  f. = 0
  Do i = 1 to length(c)
    d = substr(c,i,1)
    f.d = f.d + 1
  End
  Do i = 1 to length(p)
    d = substr(p,i,1)
    If f.d ¬= 0 then Return
    f.d = f.d - 1
  End
  Do i = 0 to 9
    If f.i ¬= 0 then Return
  End
  'output' x y p
Return
```

C Program to Search for Vampire Numbers

James Allen from Mountain View, California, searched for vampires using a C program.

```
/*
 * It's hard to do arithmetic bigger than "long" with C;
 * let's just look for the largest vampire that fits in a long,
 * by printing successively better candidates.
 */
typedef  unsigned long u_long;

main()
{
    u_long    a, b;
    u_long    bound= 0xffffffff;
    u_long    best = 4200000000;

    for (a = 0xffff; a > 0; a—)
        for (b = bound / a; b >= a && b >= best / a; b—)
            if (isvampire(a, b))
                printf("%5lu %5lu %7lu n",
                    a, b, best = a * b);
    exit(0);
}

isvampire(a, b)
    u_long    a, b;
{
    u_long    t;
    char table[11];

    strcpy(table, "          ");
    for (t = a; t; t /= 10)
        ++table[t % 10];
    for (t = b; t; t /= 10)
        ++table[t % 10];
    for (t=a*b; t; t /= 10)
        —table[t % 10];
    return !strcmp(table, "          ");
}
```

BASIC Program to Search for Vampires

```
' vampire- vampire number search...
PRINT "Begin; vampire..."
size = 3            ' 2 three-digit numbers
yup = 1
nope = -1
OPEN "a:six" FOR OUTPUT ACCESS WRITE AS #1
' The ¬ symbol means exponentiation

FOR i = 10 ¬ (size - 1) TO (10 ¬ size - 1)  ' first number
```

```
FOR j - i TO 10 ¬ size - 1' second number
   mult - i * j
   ich - i
   jch - j
      FOR k - 1 TO size * 2   ' set up search array...
         search(k) - mult MOD 10
         mult - (mult - (mult MOD 10)) / 10
         NEXT
      FOR k - size * 2 TO 1 STEP -1' fix leading zeros...
         IF search(k) - 0 THEN search(k) - 99 ELSE k - -99
         NEXT
      FOR m - 1 TO size   ' search array for first digits...
         check - ich MOD 10
         ich - (ich - (ich MOD 10)) / 10
         FOR k - 1 TO size * 2   ' find only first number match...
            IF check - search(k) THEN
               search(k) - -1
               k - 100
               END IF
            NEXT
         NEXT
      FOR m - 1 TO size       ' search array for second digits...
         check - jch MOD 10
         jch - (jch - (jch MOD 10)) / 10
         FOR k - 1 TO size * 2   '
            IF check - search(k) THEN
               search(k) - -1
               k - 100
               END IF
            NEXT
         NEXT
   match - yup
   FOR m - 1 TO size * 2
      IF search(m) >- 0 THEN
         match - nope
         m - 100
         END IF
      NEXT
   IF match - yup THEN
      count - count + 1
      PRINT count; i; j; i * j;
      PRINT #1, count; i; j; i * j
      END IF
   NEXT
NEXT
CLOSE
```

C Program to Search for Vampire Numbers

```
/* version 1.0
 * code generates numbers like 80*86-6880 called vampire numbers.
 */
/* Bugs: Truncation error in vamp_size(argument entered) calculations
 * where the argument is odd, gives a few extra vampires.  It will
```

```
 * calculate the six-digit vamps before moving on to seven, etc.
This is
 * an unfortunate behavior that I don't wish to fix at this time.
 *
 * This code ignores vampires whose components do not equal in size.
 * For example, 3 * 510 = 1503 is not considered a vampire by this
 * method.  I am not sure whether these are true vampires or not, but
 * I don't wish to find them here.
 */
/* Written by John A. Cairns  for C. Pickover */
/* Feature test switches */
#define _POSIX_SOURCE 1        /* It's POSIX, but it should compile
                                * on most ANSI C compilers also.
                                */

/* System headers */
#include <stdio.h>
#include <stdlib.h>
#include <errno.h>
#include <string.h>
#include <math.h>
#include <limits.h>

/* Structures and unions */
typedef struct /* struct containing the pertinent data */
{
   /* huge struct, with numbers of digits contained in
    * the number we are exploring.
    */
   unsigned int zero;
   unsigned int one;
   unsigned int two;
   unsigned int three;        /* number of times this number occurs */
   unsigned int four;
   unsigned int five;
   unsigned int six;
   unsigned int seven;
   unsigned int eight;
   unsigned int nine;

   char * intstring;    /* a string containing 'value' */

   unsigned int value;  /* the number with info above */
} decimal;


/* Functions */

/* declaration */

char * itos(unsigned int n);
/* take n and store it in a string */

void countdigit(decimal * number);
```

```
/* count the digits */

decimal * mult_count(decimal * x, decimal * y);
/* multiply x and y and return the
 * value.  Also counts digits.
 */

char * itos(unsigned int n)
{
  /* return a pointer to a string */

  static char * sint; /* string int */
  unsigned int size; /* compute the size here */

  size = (unsigned int)log10(n) + 2;  /* the size of the string
        + NULL */
  sint = (char *)malloc(sizeof(char) * size);
  if (sprintf(sint, "%-u", n) != (size - 1))
    {
      perror("Unable to fit number in string ");
      /* can't do anyting if this doesn't work */
      abort();
    }

  return sint;
}

void countdigit(decimal * number)
{
  /* count the number of each digit and place appropriately */
  int i;

  number->intstring = itos(number->value);

  number->nine = number->eight = number->seven = number->six =
    number->five = number->four = number->three = number->two =
      number->one = number->zero = 0;
  /* all equal zero at start! */

  for (i=0; i<strlen(number->intstring); i++)
    {
      switch (number->intstring[i])
      {
      case '0' : number->zero++;
      case '1' : number->one++;
      case '2' : number->two++;
      case '3' : number->three++;
      case '4' : number->four++;
      case '5' : number->five++;
      case '6' : number->six++;
      case '7' : number->seven++;
      case '8' : number->eight++;
      case '9' : number->nine++;
```

```
    }
  }

  /* free memory used by the string */
  (void)free(number->intstring);

  return;
}


decimal * mult_count(decimal * x, decimal * y)
{
  /* multiply the numbers and compute their digits */

  static decimal ans;

  ans.value = x->value * y->value;
  countdigit(x);
  countdigit(y);
  countdigit(&ans);

  return &ans; /* return a pointer to the answer */
}

/* Main */

int main(int argc, char *argv[])

{
  /* expects a vampire size
   * compute the product and test for a vampire number.
   */

  unsigned int vamp_size, top, bottom, i, j; /* place holders */
  decimal * product;              /* result of mult_count */
  decimal multiplicand1, multiplicand2; /* the values of interest */

  if (argc == 2)
    {
      (void)sscanf(argv[1], "%u", &vamp_size);
      if (vamp_size == 1 || vamp_size == 0) vamp_size = 2;
                    /* allow these values, but
                     * do not allow vamp_size
                     * to equal one or zero.
                     * 0, 1, and 2 don't yield
                     * vampires anyway.
                     */
      if (vamp_size > (unsigned int)log10(UINT_MAX))
      {
        fprintf(stderr, " nUsage: %s <size of vampires> n"
            "Size must be less than %u and more than 0. n n",
            argv[0], (unsigned int)log10(UINT_MAX) + 1);
        exit(0);
```

```
};
}
else
{
   fprintf(stderr, " nUsage: %s <size of vampires> n n",
       argv[0]);
   exit(0);
};


top = (unsigned int)pow(10.0, (double)vamp_size/2.0);
                         /* maximum to multiply by
                          * where the result is
                          * the desired number of
                          * digits.
                          */
bottom = (unsigned int)(pow(10.0, (double)vamp_size/2.0)/10.0);
                         /* minimum to multiply by
                          * where the result is
                          * the desired number of
                          * digits.
                          */

/* two nested for loops to find all the numbers */

for (i=bottom; i<=top; i++)
  {
    for (j=i; j<=top; j++)
  {
    /* go to town */

    multiplicand1.value = i;
    multiplicand2.value = j;
    product = mult_count(&multiplicand1, &multiplicand2);

    /* the test for a vampire is serious, but here it is. */

    if (
       (product->zero == multiplicand1.zero + multiplicand2.zero) &&
       (product->one == multiplicand1.one + multiplicand2.one) &&
       (product->two == multiplicand1.two + multiplicand2.two) &&
       (product->three == multiplicand1.three + multiplicand2.
           three) &&
       (product->four == multiplicand1.four + multiplicand2.four) &&
       (product->five == multiplicand1.five + multiplicand2.five) &&
       (product->six == multiplicand1.six + multiplicand2.six) &&
       (product->seven == multiplicand1.seven + multiplicand2.
           seven) &&
       (product->eight == multiplicand1.eight + multiplicand2.
           eight) &&
       (product->nine == multiplicand1.nine + multiplicand2.nine))

       {
         (void)printf("%-4u x %4u  = %8u\n", multiplicand1.value,
             multiplicand2.value, product->value);
```

```
      }
    }
  }
  exit(0);
}
```

QuickBASIC Program to Search for Vampire Numbers

Steve Frye from North Carolina used the following BAISC program (BASICA or QuickBASIC) to search for vampire numbers.

```
1 '/* Searching for Vampire numbers */
5 outfile$ = "d: vampire.num"        'Output file name
10 MaxNbrOfVampires% = 1000           'Size of array to hold vampires
15 ProgressInterval% = 10             'How often to print a progress
                                       indicator
500 DEFDBL A-Z
520 DIM vampires(MaxNbrOfVampires%)
570 VampireCount = 0
1000 CLS
1001 PRINT "Search for Vampires!"
1002 INPUT ; "Starting at: ", starting
1003 PRINT " "
1005 IF starting = 0 THEN starting = 1000
1010 INPUT ; "Ending at: ", ending
1011 PRINT " "
1015 IF ending = 0 THEN ending = starting * 100 - 1
1020 IF ending >= 1000 THEN GOTO 1025
1022 PRINT "There are no vampires under 1000"
1023 GOTO 1399
1025 MinVampiredigits% = LEN(STR$(starting)) - 1
1030 IF MinVampiredigits% MOD 2 = 1 THEN MinVampiredigits% =
     MinVampiredigits% + 1
1033 MaxVampireDigits% = LEN(STR$(ending)) - 1
1035 IF MaxVampireDigits% MOD 2 = 1 THEN MaxVampireDigits% =
     MaxVampireDigits% - 1
1040 OPEN outfile$ FOR OUTPUT AS #1
1045 PRINT #1, "Searching for vampires greater than "; starting;
     " and less than or equal to "; ending
1050 FOR VampireDigits% = MinVampiredigits% TO MaxVampireDigits%
     STEP 2
1052    'Next line uses exponentiation
1055    TargetRangeMin = 10 ^ (VampireDigits% - 1)
1060    IF TargetRangeMin < starting THEN TargetRangeMin = starting
1065    TargetRangeMax = 10 * TargetRangeMin - 1
1070    IF TargetRangeMax > ending THEN TargetRangeMax = ending
1073    'Next line uses exponentiation
1075    BiggestFactor = (10 ^ (VampireDigits% / 2)) - 1
1080    InitialFirstFactor = int(TargetRangeMin / BiggestFactor)
1085    FinalFirstFactor = BiggestFactor
1100    PRINT #1, "Looking in the range "; TargetRangeMin; " to ";
        TargetRangeMax
```

```
1105    PRINT "Looking in the range "; TargetRangeMin; " to ";
        TargetRangeMax
1110    PRINT "First factor will range from "; InitialFirstFactor;
        " to "; FinalFirstFactor
1115    FOR i! = InitialFirstFactor TO FinalFirstFactor
1117      IF i! MOD ProgressInterval% = 0 THEN PRINT i!
1120      FOR j! = i! TO BiggestFactor
1125        product = i! * j!
1130        IF product < TargetRangeMin THEN GOTO 1190
1135        IF product > TargetRangeMax THEN GOTO 1195
1140        GOSUB 5000    'see if product is a vampire
1145        IF vampire$ = "no" THEN GOTO 1190
1147        GOSUB 6000    'see if the vampire is a duplicate, ie,
        more than one factorization
1155        PRINT #1, i!; " x "; j!; " = "; product;
1160        PRINT SPACE$(VampireDigits% / 2 + 2); i!; " x "; j!;
        " = "; product;
1165        IF duplicate$ = "yes" THEN GOTO 1180
1170        PRINT #1, " "
1175        PRINT " "
1177        GOTO 1190
1180        PRINT #1, " duplicate"
1185        PRINT " duplicate"
1190      NEXT j!
1195    NEXT i!
1200 NEXT VampireDigits%
1300 PRINT " "
1305 PRINT "There were "; VampireCount; " distinct vampire numbers
     found."
1395 CLOSE #1
1399 END


5000 '*************************************************************
5001 '* test for vampireness                                     *
5002 '*************************************************************
5004 teststring$ = RIGHT$(STR$(product), VampireDigits%)
5005 FactorLength% = VampireDigits% / 2
5006 first$ = RIGHT$(STR$(i!), FactorLength%)
5007 second$ = RIGHT$(STR$(j!), FactorLength%)
5008 vampire$ = "yes"
5010 FOR ii% = 1 TO FactorLength%
5015   ptr% = INSTR(1, teststring$, MID$(first$, ii%, 1))
5020   IF ptr% = 0 THEN GOTO 5100
5025   MID$(teststring$, ptr%, 1) = " "
5030   ptr% = INSTR(1, teststring$, MID$(second$, ii%, 1))
5035   IF ptr% = 0 THEN GOTO 5100
5040   MID$(teststring$, ptr%, 1) = " "
5060 NEXT ii%
5070 GOTO 5999
5100 vampire$ = "no"
5999 RETURN
6000 '*************************************************************
6001 '* test for duplicate vampires                              *
6002 '*************************************************************
```

```
6003 duplicate$ = "no"
6005 FOR ii! = 1 TO VampireCount
6010   IF vampires(ii!) = product THEN GOTO 6700
6200 NEXT ii!
6210 VampireCount = VampireCount + 1
6220 vampires(VampireCount) = product
6300 GOTO 6999
6700 duplicate$ = "yes"
6999 RETURN
```

All Six-Digit Vampires

The following is a list of all six-digit vampire numbers.

$201 \times 510 = 102{,}510$	$260 \times 401 = 104{,}260$	$210 \times 501 = 105{,}210$
$204 \times 516 = 105{,}264$	$150 \times 705 = 105{,}750$	$135 \times 801 = 108{,}135$
$158 \times 701 = 110{,}758$	$152 \times 761 = 115{,}672$	$161 \times 725 = 116{,}725$
$167 \times 701 = 117{,}067$	$141 \times 840 = 118{,}440$	$201 \times 600 = 120{,}600$
$231 \times 534 = 123{,}354$	$281 \times 443 = 124{,}483$	$152 \times 824 = 125{,}248$
$231 \times 543 = 125{,}433$	$204 \times 615 = 125{,}460$	$246 \times 510 = 125{,}460$
$251 \times 500 = 125{,}500$	$210 \times 600 = 126{,}000$	$201 \times 627 = 126{,}027$
$261 \times 486 = 126{,}846$	$140 \times 926 = 129{,}640$	$179 \times 725 = 129{,}775$
$311 \times 422 = 131{,}242$	$323 \times 410 = 132{,}430$	$315 \times 423 = 133{,}245$
$317 \times 425 = 134{,}725$	$231 \times 588 = 135{,}828$	$351 \times 387 = 135{,}837$
$215 \times 635 = 136{,}525$	$146 \times 938 = 136{,}948$	$150 \times 930 = 139{,}500$
$350 \times 401 = 140{,}350$	$350 \times 410 = 143{,}500$	$351 \times 414 = 145{,}314$
$317 \times 461 = 146{,}137$	$156 \times 942 = 146{,}952$	$300 \times 501 = 150{,}300$
$251 \times 608 = 152{,}608$	$261 \times 585 = 152{,}685$	$300 \times 510 = 153{,}000$
$356 \times 431 = 153{,}436$	$240 \times 651 = 156{,}240$	$269 \times 581 = 156{,}289$
$165 \times 951 = 156{,}915$	$176 \times 926 = 162{,}976$	$396 \times 414 = 163{,}944$
$221 \times 782 = 172{,}822$	$231 \times 750 = 173{,}250$	$371 \times 470 = 174{,}370$
$231 \times 759 = 175{,}329$	$225 \times 801 = 180{,}225$	$201 \times 897 = 180{,}297$
$225 \times 810 = 182{,}250$	$281 \times 650 = 182{,}650$	$210 \times 870 = 182{,}700$
$216 \times 864 = 186{,}624$	$210 \times 906 = 190{,}260$	$210 \times 915 = 192{,}150$
$327 \times 591 = 193{,}257$	$395 \times 491 = 193{,}945$	$275 \times 719 = 197{,}725$
$252 \times 801 = 201{,}852$	$255 \times 807 = 205{,}785$	$216 \times 981 = 211{,}896$
$341 \times 626 = 213{,}466$	$251 \times 860 = 215{,}860$	$323 \times 671 = 216{,}733$
$321 \times 678 = 217{,}638$	$248 \times 881 = 218{,}488$	$270 \times 810 = 218{,}700$
$269 \times 842 = 226{,}498$	$276 \times 822 = 226{,}872$	$248 \times 926 = 229{,}648$
$338 \times 692 = 233{,}896$	$461 \times 524 = 241{,}564$	$422 \times 581 = 245{,}182$
$296 \times 851 = 251{,}896$	$350 \times 725 = 253{,}750$	$470 \times 542 = 254{,}740$
$323 \times 806 = 260{,}338$	$284 \times 926 = 262{,}984$	$437 \times 602 = 263{,}074$
$489 \times 582 = 284{,}598$	$420 \times 678 = 284{,}760$	$468 \times 612 = 286{,}416$
$320 \times 926 = 296{,}320$	$431 \times 707 = 304{,}717$	$431 \times 725 = 312{,}475$
$321 \times 975 = 312{,}975$	$534 \times 591 = 315{,}594$	$351 \times 900 = 315{,}900$
$351 \times 909 = 319{,}059$	$336 \times 951 = 319{,}536$	$524 \times 623 = 326{,}452$
$342 \times 963 = 329{,}346$	$356 \times 926 = 329{,}656$	$530 \times 635 = 336{,}550$

360 × 936 = 336,960	392 × 863 = 338,296	533 × 641 = 341,653
366 × 948 = 346,968	369 × 981 = 361,989	392 × 926 = 362,992
533 × 686 = 365,638	585 × 630 = 368,550	381 × 969 = 369,189
383 × 971 = 371,893	473 × 800 = 378,400	431 × 878 = 378,418
435 × 870 = 378,450	432 × 891 = 384,912	465 × 831 = 386,415
593 × 662 = 392,566	446 × 908 = 404,968	491 × 845 = 414,895
641 × 650 = 416,650	468 × 891 = 416,988	482 × 890 = 428,980
464 × 926 = 429,664	476 × 941 = 447,916	540 × 846 = 456,840
650 × 704 = 457,600	546 × 840 = 458,640	570 × 834 = 475,380
624 × 780 = 486,720	549 × 891 = 489,159	545 × 899 = 489,955
590 × 845 = 498,550	681 × 759 = 516,879	572 × 926 = 529,672
563 × 953 = 536,539	630 × 855 = 538,650	588 × 951 = 559,188
657 × 864 = 567,648	650 × 875 = 568,750	680 × 926 = 629,680
650 × 983 = 638,950	720 × 936 = 673,920	750 × 906 = 679,500
800 × 860 = 688,000	788 × 926 = 729,688	765 × 963 = 736,695
843 × 876 = 738,468	776 × 992 = 769,792	875 × 902 = 789,250
825 × 957 = 789,525	855 × 927 = 792,585	807 × 984 = 794,088
891 × 909 = 809,919	894 × 906 = 809,964	858 × 951 = 815,958
896 × 926 = 829,696	891 × 945 = 841,995	953 × 986 = 939,658

C Program to Solve Ontario Puzzle

Tomas Oliveira e Silva from Portugal wrote the following C code to solve the Ontario puzzle.

```
/*
** Let the digits, not necessarily all different, of the first row be
** Row 1: p[0],p[1],....,p[N-1]. Let Row 2: q[0],...,q[N-1] be one
   solution
** of the problem stated above. Then one solution of the problem
** Row 1: P[p[0],....,p[N-1]], where P[] represents a permutation of
   its
** arguments, is Row 2: P[q[0],...,q[N-1]]. Therefore, to generate all
** interesting cases without duplications one has to consider only
   the cases
** where p[0] <= p[1] <= ... <= p[N-1]. It turns out that there are
   only
** 92378 such cases for N=10 (compare with 10^10 in a more naive
   approach to
** this problem). An exhaustive search is therefore possible.
**
** To solve the contest we generate all solutions for all 92378
   cases, sort
** the solutions in increasing order (i.e., such that q[0] <= ...
   <= q[N-1]),
** and construct a directed graph connecting two nodes only if the
   second is
** a solution for the row represented by the first. This requires
   about 4
```

```
** hours on a DEC Alpha 3000 400 and about 7M bytes of memory.
   Finally, we
** examine the graph looking for large "paths" and, why not, large
   loops.
*/

#include <math.h>
#include <stdio.h>
#include <stdlib.h>

#ifndef N
#  define N       10
#endif

#if N == 10
#  define nNodes 92378
#else
#  define nNodes 10000
#endif

#define maxSol    100
#define maxPath   100

typedef struct node
{
  char p[N];        /* Pattern */
  int nSol;         /* Number of solutions */
  struct node **sol; /* List of solutions */
}
node;

static node nodes[nNodes];


static int findSolutions(int *p,int **sol)
{
  /*
  ** z[i] contains the number of times the digit i occurs
  ** y[] contains the number of times each digit occurs is z[]
  ** p[] contains the "pattern" of digits of the "key"
  */
  int z[N],y[N + 1];
  int i,j,nSol;

  /*** Initialization ***/

  z[0] = N;
  for(j = 1;j < N;j++)
    z[j] = 0;
  i = 0;
  nSol = 0;

  /*** Exhaustive search ***/

  for(;;)
```

```
{ /*** The number of times this loop is executed is nNodes ***/
  /*** Test current number ***/
  for(j = 0;j <= N;j++)
    y[j] = 0;
  for(j = 0;j < N;j++)
    y[z[p[j]]]++;
  for(j = 0;j < N && z[j] == y[j];j++)
    ;
  if(j == N)
  { /*** Bingo! ***/
    if(nSol == maxSol)
    {
      fprintf(stderr,"Increase maxSol n");
      exit(1);
    }
    for(j = 0;j < N;j++)
      sol[nSol][j] = z[p[j]];
    nSol++;
  }

  /*** Find next possible number ***/
  if(i == N - 1)
  { /*** Go back ***/
    j = z[i];
    z[i] = 0;
    while(--i >= 0 && z[i] == 0)
      ;
    if(i < 0)
      break; /*** Exhaustive search is finished ***/
    z[i]--;
    z[++i] = j + 1;
  }
  else
  { /*** Go forward ***/
    z[i]--;
    z[++i] = 1;
  }
}
return(nSol);
}


static void initializeNodes(void)
{
  int z[N];
  int i,n;

  /*** Nodes generated in lexicographic order ***/
  n = 0;
  for(i = 0;i < N;i++)
    z[i] = 0;
  for(;;)
  {
    /*** Save current node ***/
```

```
    if(n >= nNodes)
    {
      fprintf(stderr,"Increase nNodes n");
      exit(1);
    }
    for(i = 0;i < N;i++)
      nodes[n].p[i] = (char)z[i];
    nodes[n].nSol = 0;
    n++;
/***************************
    for(i = 0;i < N;i++)
      printf("%d",z[i]);
    printf(" n");
***************************/

    /*** Find next node ***/
    for(i = N - 1;i >= 0 && z[i] == N - 1;-i)
      ;
    if(i < 0)
      break;
    z[i]++;
    while(++i < N)
      z[i] = z[i - 1];
  }
  if(n < nNodes)
  {
    fprintf(stderr,"Set nNodes to %d n",n);
    exit(1);
  }
}


static int compareNodes(node *n1,node *n2)
{ /*** This code could be replaced by strncmp ***/
  int i;
  for(i = 0;i < N && n1->p[i] == n2->p[i];i++)
    ;
  if(i == N)
    return(0);
  return(n1->p[i] < n2->p[i] ? -1 : 1);
}

static node *findNode(int *p)
{
  int i,j,k;
  node n;

  /*** Sort p[] ***/
  for(j = N - 1;j > 0;-j)
    for(i = 0;i < j;i++)
      if(p[i] > p[i + 1])
      {
        k = p[i];
        p[i] = p[i + 1];
```

```
        p[i + 1] = k;
    }

  /*** Search for the node having this p[] ***/
  for(i = 0;i < N;i++)
    n.p[i] = (char)p[i];
  return((node *)bsearch(&n,nodes,nNodes,sizeof(node),
    compareNodes));
}
static void findAllSolutions(void)
{
  int i,j,n,p[N],*sol[maxSol],solData[maxSol][N];

  for(i = 0;i < maxSol;i++)
    sol[i] = solData[i];

  for(i = 0;i < nNodes;i++)
  {
    for(j = 0;j < N;j++)
      p[j] = (int)nodes[i].p[j];
    n = findSolutions(p,sol);
    nodes[i].nSol = n;
    if(n > 0)
    {
      nodes[i].sol = (node **)malloc(n * sizeof(node *));
      if(nodes[i].sol == NULL)
      {
        fprintf(stderr,"Out of memory n");
        exit(1);
      }
      for(j = 0;j < n;j++)
      {
        nodes[i].sol[j] = findNode(sol[j]);
        if(nodes[i].sol[j] == NULL)
        {
          fprintf(stderr,"Impossible n");
          exit(1);
        }
      }
    }
  }
}


static int nLoop;

static node *lLoop[maxPath];
static int pLoopSize,lLoopSize;

static node *bestPath[maxPath];
static int bestPathSize,bestPathHeaderSize;

static void registerLoop(node **path,int pSize,int lSize)
{
```

```
  int i;

  nLoop++;

  if(path[0] == path[pSize - 1] && lSize > lLoopSize)
  {
    pLoopSize = pSize;
    for(i = 0;i < pSize;i++)
      lLoop[i] = path[i];
    lLoopSize = lSize;
  }

  if(pSize - lSize > bestPathHeaderSize)
  {
    bestPathSize = pSize;
    for(i = 0;i < pSize;i++)
      bestPath[i] = path[i];
    bestPathHeaderSize = pSize - lSize;
  }
}


static int nsPath;

static node *sPath[maxPath];
static int sPathSize;
static void registerPath(node **path,int pSize)
{
  int i;

  nsPath++;

  if(pSize > sPathSize)
  {
    sPathSize = pSize;
    for(i = 0;i < pSize;i++)
      sPath[i] = path[i];
  }
}

static void followPath(node **path,int pSize)
{
  int i,j,k;
  node *n;

  if(pSize == maxPath)
  {
    fprintf(stderr,"Increase maxPath n");
    exit(1);
  }
  k = path[pSize - 1]->nSol;
  if(k == 0)
    registerPath(path,pSize);
  else
```

```
      for(i - 0;i < k;i++)
      {
        n - path[pSize - 1]->sol[i];
        path[pSize] - n;
        for(j - pSize - 1;j >- 0 && path[j] !- n;-j)
          ;
        if(j < 0)
          followPath(path,pSize + 1);
        else
          registerLoop(path,pSize + 1,pSize - j);
      }
}


static void printPath(char *header,node **path,int pSize)
{
  int i,j;

  for(i - 0;i < pSize;i++)
  {
    printf("%s",header);
    for(j - 0;j < N;j++)
      printf("%d",path[i]->p[j]);
    printf("\n");
  }
}

static void generateAllPaths(void)
{
  node *path[maxPath];
  int i,j;

  nLoop - 0;
  lLoopSize - 0;
  bestPathHeaderSize - 0;

  nsPath - 0;
  sPathSize - 0;;

  for(i - 0;i < nNodes;i++)
  {
    path[0] - nodes + i;
    followPath(path,1);
  }

  printf("\n/*\n** Program output for N - %d\n**\n**\n",N);
  printf("** Paths with loops: %d\n** n",nLoop);
  printf("**    Largest loop: %d\n",lLoopSize);
  printPath("**      ",lLoop,pLoopSize);
  printf("** n**   \"Best\" path: %d\n",bestPathSize);
  printPath("**      ",bestPath,bestPathSize);
  printf("**\n**\n** Paths without loops: %d\n**\n",nsPath);
  printf("**    Longest path: %d n",sPathSize);
  printPath("**      ",sPath,sPathSize);
  i - 0;
```

```
    for(j = 1;j < nNodes;j++)
      if(nodes[j].nSol > nodes[i].nSol)
        i = j;
    printf("**\n** Highest number of solutions: %d\n",nodes[i].nSol);
    path[0] = nodes + i;
    printPath("**    ",path,1);
    printPath("**      ",nodes[i].sol,nodes[i].nSol);
    printf("*/\n");
}


static void solveCase(char *pattern)
{
    int i,j,n,p[N],*sol[maxSol],solData[maxSol][N];

    for(j = 0;j < N;j++)
    {
      p[j] = pattern[j] - '0';
      if(p[j] < 0 || p[j] >= N)
        exit(1);
    }

    for(j = 0;j < N;j++)
      printf("%d",p[j]);
    printf("\n");
    for(j = 0;j < N;j++)
      printf("=");
    printf("\n");

    for(i = 0;i < maxSol;i++)
      sol[i] = solData[i];
    n = findSolutions(p,sol);
    for(i = 0;i < n;i++)
    {
      for(j = 0;j < N;j++)
        printf("%d",sol[i][j]);
      printf("\n");
    }
    printf("\n");
}


int main(int argc,char **argv)
{
    if(argc == 1)
    {
      initializeNodes();
      findAllSolutions();
      generateAllPaths();
    }
    else if(argc > 2)
    {
      fprintf(stderr,"Usage: %s [pattern]",argv[0]);
      return(1);
    }
```

```
    else
      solveCase(argv[1]);
    return(0);
}

/*
** Program output for N = 10
**
**
** Paths with loops: 125502
**
**    Largest loop: 4
**      0000111244  0111133444  0001122223  0022233334  0000111244
**
**    "Best" path: 9
**      0000122245  0022223444  0001111233  0111333344  0000112224
**      0022224444  0000000088  0022222222  0000000088
**
** Paths without loops: 91450
**
**    Longest path: 5
**      1223334444  1111223334  0000022233  0003333355  0000055555
**
** Highest number of solutions: 43
1223334444 0000000000 0000000001 0000000022 0000000112 0000000122
0000000333 0000001113 0000001333 0000004444 0000011114 0000014444
0000022233 0000022333 0000111223 0000112333 0000122233 0000122333
0000222244 0000224444 0001111224 0001124444 0001222244 0001224444
0003333444 0003334444 0011113334 0011134444 0013333444 0013334444
0222233344 0222334444 0223333444 0223334444 1111222334 1111223334
1112222344 1112234444 1123333444 1123334444 1222233444 1222334444
1223333444 1223334444
*/
```

C Program to Solve Cyclotron Puzzles

The following C code by Heiner Marxen solves the various cyclotron puzzles.

```c
/*
 * Solve Cyclotron Puzzle, as posed by Cliff Pickover
 * Author: Heiner Marxen
 */
#include <stdio.h>
#include <stdlib.h>
#include <string.h>              /* memset, strspn */
#define NX      20              /* logical horizontal length */
#define NY      15              /* logical vertical length */
#define MX      (NX+2)          /* physical horizontal length */
#define MY      (NY+2)          /* physical vertical length */
#define MXY     (MX*MY)         /* physical size */
#define MAXV    101
#if 0                           /* CHECK for your machine */
    typedef char    Val; /* must hold -MAXV..+MAXV */
```

```
#else
    typedef short  Val; /* must hold -MAXV..+MAXV */
#endif

typedef unsigned long   Counter;

#define PERC(part, tot) (  (tot)             \
                ? (100.0 * (((double)(part)) / ((double)(tot)))) \
                : 0.0                \
                )
#define FOR(v,mx) for( (v)=0 ; (v)<(mx) ; ++(v) )

static int    optv = 0;      /* verbose level */
static int    optt = 0;      /* trace level */
static int    optb = 0;      /* better show level */
static int    opts = 0;      /* stats level */
static int    opti[2]  = { 0, 0 };    /* y, x */
static int    optC = 0;      /* cut level */

    static void
show_val( v )
    int  v;
{
    if( v ) {
    if( v <= -100 ) {
        printf("-%02d", v+100);
    }else {
        printf("%3d", v);
    }
    }else {
    printf("%3s", "");
    }
}

    static void
show_maze_path( maze, npath, xpath, ypath )
    Val   maze [MY][MX];
    Val   xpath[MAXV];
    Val   ypath[MAXV];
{
    Val   path [MY][MX];
    int       y,x,v;
    FOR(y,MY) FOR(x,MX) path[y][x] = 0;
    FOR(v,npath) path[ypath[v]][xpath[v]] = v+1;
    FOR(y,MY) {
    FOR(x,MX) {
        v = maze[y][x];
        if( v && path[y][x] ) v = -v;
        show_val(v);
    }
    printf("\n");
    }
    FOR(y,MY) {
    FOR(x,MX) {
```

```
        v = maze[y][x];
        if( v ) {
         if( path[y][x] ) {
             printf("%3d", path[y][x]);
         }else {
             printf("%3s", "-");
         }
        }else {
         printf("%3s", "");
        }
    }
    printf("\n");
    }
}

    static void
show_maze( maze )
    Val   maze[MY][MX];
{
    int      y,x;
    FOR(y,MY) {
    FOR(x,MX) show_val((int) maze[y][x]);
    printf("\n");
    }
}

static int     vfreq[MAXV];
static int     vocc;
static int     vtot;
    static void
check_maze( maze )
    Val   maze[MY][MX];
{
    int      y,x,v;
    FOR(y,MY) {
    FOR(x,MX) {
        v = maze[y][x];
        if( (v < 0) || (v >= MAXV) ) {
         fprintf(stderr, "maze[%d][%d] = %d out of range\n", y, x, v);
         exit(1);
        }
        if( v ) {
         if( x==0 || x==(MX-1) || y==0 || y==(MY-1) ) {
             fprintf(stderr, "maze[%d][%d] = %d not border\n",
             y, x, v);
             exit(1);
         }
        }
    }
    }
    vocc = 0;  vtot = 0;
    FOR(v,MAXV) vfreq[v] = 0;
    FOR(y,MY) FOR(x,MX) vfreq[maze[y][x]] += 1;
    FOR(v,MAXV) vocc += !! vfreq[v];
    FOR(v,MAXV) vtot +=    vfreq[v];
```

```
    vocc -= !! vfreq[0];
    vtot -=    vfreq[0];
    if( optv > 1 ) {
    FOR(v,MAXV) {
        printf("val %3d: %5d times\n", v, vfreq[v]);
    }
    }
    printf("Total      nonzero values: %3d\n", vtot);
    printf("Different nonzero values: %3d\n", vocc);
}

#define DIRS   4
static short  xstep[DIRS] ={ 1,  0, -1,  0 };
static short  ystep[DIRS] ={ 0,  1,  0, -1 };
static char   vbad[MAXV];
static Val    xpath[MAXV];
static Val    ypath[MAXV];
static int    npath;
static int    b_npath;
static int    b_valid;
static Val    b_xpath[MAXV];
static Val    b_ypath[MAXV];
static Counter     b_optim;

static Counter     cntdep[MAXV];
static Counter     cntcut[MAXV];         /* [dep] */
static Counter     cntcan[MAXV];         /* [reach] */
static Counter     cntuse[MAXV];         /* [reach] */

    static void
show_poss( maze )
    Val       maze[MY][MX];
{
    Val       poss[MY][MX];
    int       x, y, v;

    FOR(y,MY) FOR(x,MX) poss[y][x] = maze[y][x];
    FOR(v,npath) {
    x = xpath[v];
    y = ypath[v];
    poss[y][x] = - poss[y][x];
    }
    FOR(y,MY) FOR(x,MX) {
    v = poss[y][x];
    if( (v > 0) && vbad[v] ) {
        poss[y][x] = 0;
    }
    }
    show_maze(poss);
}

    static void
better_maze( maze )
    Val       maze[MY][MX];
{
```

```
    int        v;

    if( b_npath >= vocc ) {
    b_optim += 1;
    }
    if( optb > 0 ) {
    printf("Better maze path length %3d\n", npath);
    }
    if( optb > 1 ) {
    show_maze_path(maze, npath, xpath, ypath);
    if( optv > 0 ) {
        FOR(v,npath) {
         printf("[%2d/%2d]", ypath[v], xpath[v]);
         if( (v+1)%10 == 0 ) printf("\n");
         }
         if( v%10 != 0 ) printf("\n");
    }
    }
    if( optb > 2 ) {
    show_poss(maze);
    }
    if( optb > 0 ) { (void) fflush(stdout); }
    FOR(v,npath) {
    b_xpath[v] = xpath[v];
    b_ypath[v] = ypath[v];
    }
    b_npath = npath; b_valid = 1;
}

    static void
tell_rec( y, x, v, c )
    int        y,x,v,c;
{
    int        n;
    n = npath;
    printf("%3d ", n);
    for( ; n > 10 ; n-=10 ) {
    printf("X");
    }
    for( ; n > 0 ; n-=1 ) {
    printf("%c", c);
    }
    printf("\t%2d/%2d = %2d", y, x, v);
    printf("\n");
    (void) fflush(stdout);
}

    static int          /* maximum possible path extension */
cnt_reach( maze, geo, y0, x0, rlim )
    Val        maze[MY][MX];
    Val        geo [MY][MX];
    int        y0, x0;
    int        rlim;    /* larger results are not better */
{
    char vdid[MAXV];
```

```
    char xx[MXY];
    char yy[MXY];
    int     x, y, v, cnt, lo, hi, g, d, nx, ny;

    cnt = 0;
    v = maze[y0][x0];
    if( (v > 0) && !vbad[v] ) {
    cnt += 1;
    if( cnt >= rlim ) goto out;
    lo = 0;
    hi = 0;
    g = geo[y0][x0];

    FOR(d,DIRS) {
        if( g & (1<<d) ) {
        x = x0 + xstep[d];
        y = y0 + ystep[d];
        v = maze[y][x];
        if( (v > 0) && !vbad[v] ) {
            xx[hi] = x; yy[hi] = y;
            hi += 1;
        }
        }
    }
    if( hi ) {
#if 0
        FOR(v,MAXV) vdid[v] = 0;
#else
        memset((void*)vdid, 0, sizeof(vdid));
#endif
        v = maze[y0][x0];
        maze[y0][x0] = -v;
        vdid[v] = 1;
        for( lo=0 ; lo<hi ; ++lo ) {
        x = xx[lo];  y = yy[lo];
        v = maze[y][x];
        maze[y][x] = -v;           /* entered */
        if( ! vdid[v] ) {
            vdid[v] = 1; cnt += 1;
        }
        }
        if( cnt >= rlim ) goto back;
        for( lo=0 ; lo<hi ; ) {
        x = xx[lo];  y = yy[lo];
        ++lo;
        g = geo[y][x];
        FOR(d,DIRS) {
            if( g & (1<<d) ) {
            nx = x + xstep[d]; ny = y + ystep[d];
            v = maze[ny][nx];
            if( (v > 0) && !vbad[v] ) {
                if( ! vdid[v] ) {
                vdid[v] = 1; cnt += 1;
                if( cnt >= rlim ) goto back;
                }
```

```
                        maze[ny][nx] = -v;          /* entered */
                        xx[hi] = nx; yy[hi] = ny;
                        hi += 1;
                    }
                }
            }
        }
    back:   ;
        for( lo=hi ; —lo >= 0 ; ) {
        x = xx[lo];
        y = yy[lo];
        maze[y][x] = - maze[y][x];
        }
        maze[y0][x0] = - maze[y0][x0];
#if 0
        {
            char  mimax[2];
            char  mimay[2];
            lo = 0;
            mimax[0] = mimax[1] = x0;
            mimay[0] = mimay[1] = y0;
            while( lo < hi ) {
                x = xx[lo];
                y = yy[lo];
                ++lo;
                if( x < mimax[0] ) mimax[0] = x;else
                if( x > mimax[1] ) mimax[1] = x;
                if( y < mimay[0] ) mimay[0] = y;else
                if( y > mimay[1] ) mimay[1] = y;
            }
            printf("mima %2d-%2d * %2d-%2d = %3d [%3d]\n",
                mimax[0], mimax[1], mimay[0], mimay[1],
                (mimax[1]-mimax[0]+1)* (mimay[1]-mimay[0]+1),
                cnt
            );
        }
#endif
    }
    }
out:;
    cntcan[cnt] += 1;
    return cnt;
}

    static void
scan_maze( maze, geo, y, x )
    Val      maze[MY][MX];
    Val      geo [MY][MX];
    int      y, x;
{
    int      d, v, g, nv;
    if( b_optim ) return;
    cntdep[npath] += 1;     /* recursion depth accounting */
    v = maze[y][x];
    if( v==0 || vbad[v] ) {
```

```
    return;
    }
    ypath[npath] = y;
    xpath[npath] = x;
    npath += 1;
    vbad[v] = 1;
    {
    if( npath <= optt ) tell_rec(y,x,v, '>');
    if( npath > b_npath ) better_maze(maze);
    g = geo[y][x];
    FOR(d,DIRS) {
        if( g & (1<<d) ) {
        nv = maze[y+ystep[d]][x+xstep[d]];
        if( vbad[nv] ) {
            g &= ~(1<<d);
        }
        }
    }
    if( g ) {
        FOR(d,DIRS) {
        if( g & (1<<d) ) {
            int  rcnt;
            if( (npath <= b_npath) && (g & (g-1)) ) {
            rcnt = cnt_reach(maze, geo,
                    y+ystep[d], x+xstep[d],
                    b_npath + 1 - npath);
            }else {
            rcnt = MAXV;
            }
            if( (npath + rcnt) > b_npath ) {
            scan_maze(maze, geo, y+ystep[d], x+xstep[d]);
            }else {
            cntcut[npath] += 1;
            cntuse[rcnt ] += 1;
            }
        }
        }
    }
    if( npath <= optt ) tell_rec(y,x,v, '<');
    }
    vbad[v] = 0;
    npath -= 1;
}

    static void
start_maze( maze, geo, y, x )
    Val     maze[MY][MX];
    Val     geo [MY][MX];
    int     y,x;
{
    int     v;
    if( (optv > 0) || (optt > 0) ) {
    printf("START maze at [%2d][%2d]\n", y, x);
    (void) fflush(stdout);
    }
```

```
    FOR(v,MAXV) vbad[v] = 0;
    vbad[0] = 1;
    npath = 0;
    scan_maze(maze, geo, y, x);
    {                    /* sanity checks */
        int    errs;
        errs = 0;
        if( npath ) {
            fprintf(stderr, "Left npath = %d\n", npath);
            ++errs;
        }

        FOR(v,MAXV) {
            if( vbad[v] != (v==0) ) {
             fprintf(stderr, "Left vbad[%d] = %d\n", v, vbad[v]);
             ++errs;
            }
        }
        if( errs ) {
            exit(1);
        }
    }
}


    static void
def_geo( maze, geo )
    Val       maze[MY][MX];   /* checked, has border */
    Val       geo [MY][MX];
{
    Val       ring[MY][MX];
    int       x,y,d,nx,ny;
    FOR(y,MY) FOR(x,MX) {
    ring[y][x] = 0;
    geo [y][x] = 0;
    }
#define SETRING(r) if(maze[y][x]&&!ring[y][x]) ring[y][x] = (r);
    for( d=1 ; d<4 ; ++d ) {
    y = d      ; FOR(x,MX) SETRING(d);
    y = MY-1-d; FOR(x,MX) SETRING(d);
    x = d      ; FOR(y,MY) SETRING(d);
    x = MX-1-d; FOR(y,MY) SETRING(d);
    }
    FOR(y,MY) FOR(x,MX) {
    SETRING(9);
    }
#undef SETRING
    if( optv > 0 ) {
    show_maze(ring);
    }
    FOR(y,MY) FOR(x,MX) {
    if( ! maze[y][x] ) continue;
    FOR(d,DIRS) {
        nx = x + xstep[d];
        ny = y + ystep[d];
        if( ! maze[ny][nx] ) continue;
```

```
        if( ring[y][x] < 9 ) {
         if( ring[y][x] == ring[ny][nx] ) {
             if( (ny > y) && (x < MX/2) ) continue;
             if( (ny < y) && (x > MX/2) ) continue;
             if( (nx < x) && (y < MY/2) ) continue;
             if( (nx > x) && (y > MY/2) ) continue;
          }
         }
         geo[y][x] |= 1<<d;
     }
     }
     if( optv > 0 ) {
     show_maze(geo);
     }
}


    static void
search_maze( maze, geo )
    Val      maze[MY][MX];
    Val      geo [MY][MX];
{
    int      x, y;
    b_npath = 0;
    b_optim = 0;
    b_valid = 0;
    FOR(x,MAXV) cntdep[x] = 0;
    FOR(x,MAXV) cntcut[x] = 0;
    FOR(x,MAXV) cntcan[x] = 0;
    FOR(x,MAXV) cntuse[x] = 0;
    if( optC ) {
    b_npath = optC;
    }
    if( opti[0] || opti[1] ) {
    start_maze(maze, geo, opti[0], opti[1]);
    }else {
    FOR(y,MY) FOR(x,MX) {
        if( maze[y][x] ) {
         start_maze(maze, geo, y, x);
        }
    }
    }
    printf("BEST maze path length %3d\n", b_npath);
    if( b_valid ) {
    show_maze_path(maze, b_npath, b_xpath, b_ypath);
    }
    if( opts > 0 ) {
        Counter    tot;
        Counter    sum;
        Counter    last;
        Counter    curr;
    tot = 0;
    FOR(x,MAXV) tot += cntdep[x];
    if( tot ) {
        last = 1;
        FOR(x,MAXV) {
```

```
        curr = cntdep[x];
        if( curr || cntcut[x] ) {
            printf("dep %3d: %10lu [%6.2f%%] %7.3f %10lu\n",
                x,
                (unsigned long) curr,
                PERC(curr, tot),
                last ? (double)curr / (double)last : (double)curr,
                (unsigned long) cntcut[x]
            );
        }
        last = curr;
        }
    }
    tot = 0;
    FOR(x,MAXV) tot += cntcan[x];
    if( tot ) {
        sum = 0;
        FOR(x,MAXV) {
         curr = cntcan[x];
         if( curr || cntuse[x] ) {
             sum += curr;
             printf("can %3d: %10lu [%6.2f%%] [%6.2f%%] %10lu\n",
                x,
                (unsigned long) curr,
                PERC(curr, tot), PERC(sum , tot),
                (unsigned long) cntuse[x]
            );
         }
        }
    }
    }
    (void) fflush(stdout);
}

    static void
do_maze( maze )
    Val        maze[MY][MX];
{
    Val        geo [MY][MX];
    show_maze(maze);   check_maze(maze);
    def_geo(maze, geo);   search_maze(maze, geo);
}

    static void
do_file( fp )
    FILE*      fp;
{
    static char    lbuf[100+1];
    int        y, x, i;
    Val        maze[MY][MX];

    FOR(y,MY) FOR(x,MX) maze[y][x] = 0;
    FOR(y,NY) {
    FOR(i,100) lbuf[i] = 0;
    if( ! fgets(lbuf, 100, fp) ) {
```

```
        fprintf(stderr, "Cannot read line %d\n", y);
        exit(1);
    }
    FOR(x,NX) {
        char nbuf[4];
        FOR(i,3) nbuf[i] = lbuf[i+3*x];
        nbuf[3] = 0;
        if( nbuf[strspn(nbuf, " ")] ) {      /* has non-blank */
        maze[y+1][x+1] = 1 + atoi(nbuf);
        }
    }
    }
    do_maze(maze);
}

    static void
do_name( fname )
    char*       fname;
{
    FILE*       fp;
    fp = fopen(fname, "r");
    if( ! fp ) {
    fprintf(stderr, "Cannot open '%s' for reading\n", fname);
    exit(1);
    }
    do_file(fp);
    (void) fclose(fp);
}


main( argc, argv )
    int         argc;
    char**      argv;
{
    int         ac;
    ac = 0;
    ++ac;
    while( (ac < argc) && (argv[ac][0] == '-') ) {
        register char*  opt = argv[ac++];
    while( *++opt ) switch( *opt ) {
     case 'v': optv+= 1; continue;
     case 's': opts+= 1; continue;
     case 't': optt+= 1; continue;
     case 'T': optt+= 10; continue;
     case 'b': optb+= 1; continue;
     case 'C':
        if( ac < argc ) {
        optC = atoi(argv[ac++]);
        }
        continue;
     case 'i':
        opti[0] = opti[1] = 0;
        if( ac < argc ) {
        opti[0] = atoi(argv[ac++]);
        }
```

```
      if( ac < argc ) {
       opti[1] = atoi(argv[ac++]);
      }
      continue;
    default:
      fprintf(stderr, "Unknown option '%c'\n", *opt);
      exit(1);
    }
    }
    if( ac < argc ) {
    do_name(argv[ac++]);
    }else {
    do_file(stdin);
    }
    exit(0);
    return 1;        /* paranoia */
}
```

Notes

Preface

1. To get a feel for Usenet's popularity, consider that around 40,000 messages are sent each day to 2,000 newsgroups in Usenet. The most popular newsgroup, news.announce.newusers, has approximately 800,000 readers. A newsgroup called "alt.fan.cliffpickover" has been started for discussions on topics and problems discussed in my books.

2. Interestingly, the Internet (a major computer network linking millions of machines around the world) is growing at a phenomenal rate. Between January 1993 and January 1994, the number of connected machines grew from 1,313,000 to 2,217,000. More than 70 countries have full Internet connectivity, and about 150 have at least some electronic mail services. As I write this, there are about 20 to 25 million users of the Internet (Goodman et al., "The Global Diffusion of the Internet," *Communications of the ACM*, 37[8][1994]: 27).

Chapter 3 Infinity Machines

1. International Tesla Society, P.O. Box 5636, Colorado Springs, CO 80931.

2. An irrational number cannot be expressed as a ratio of two integers. Transcendental numbers such as $e = 2.718 \ldots$ and $\pi = 3.1415 \ldots$ and all *surds* (e.g., $\sqrt{2}$, cube roots, etc.) are irrational.

Chapter 4 Infinity World

1. Mathematicians are comfortable with the notion of different levels of infinities. For example, there are an infinity of rational numbers (fractions), but there is an even greater infinity of irrational numbers (nonrepeating, nonterminating decimals such as π). In fact, between any two infinitely close fractions dwells an infinity of irrational numbers.

2. C. Hargittai and C. Pickover, *Spiral Symmetry* (River Edge, NJ: World Scientific, 1992).

Chapter 5 Grid of the Gods

1. Various researchers have pointed out that the process of randomly selecting chords on a circle to solve this problem leads to some famous paradoxes. The term *random chord* is ambiguous unless the exact procedure for producing it is defined, as in my program code

(Appendix 1). Would our results be different if we were to randomly choose chords that were all parallel to a given direction?

2. M. Schroeder, *Number Theory in Science and Communication,* 2nd enlarged edition, 107–108 (Berlin and New York: Springer, 1990).

Chapter 6 To the Valley of the Sea Horses

1. I came across Don Webb's name in a list of members for *YLEM,* an organization consisting primarily of artists who use science and technology. In the listing, Don gave his occupation as "science-fiction writer and critic." I could not resist writing to him, and we soon tossed around the idea of collaborating on a short science-fiction tale about a rich man who becomes infatuated with fractals and infinity. This chapter is the result of our collaboration.

Don is a reviewer for the *American Book Review* and for the *New York Review of Science Fiction.* His work has appeared in more than 200 large and small magazines around the world, including all major science-fiction magazines. His story "Jesse Revenged" was selected to appear in the "Hundred Best" list in *American Short Stories 1987.* He has written several books, including *Uncle Ovid's Exercise Book,* which won the 1988 Illinois State University/Fiction Collective fiction contest. His story "Walk on the Farside" was chosen by Gardner Dozios for *The Year's Best Science Fiction XI.*

Chapter 7 The Million-Dollar, Trillion-Digit Pi Sequencing Initiative

1. The digits of transcendental numbers seem to go on forever without any rhyme, reason, or repetition. For the mathematically inclined, note that transcendental numbers cannot be expressed as the root of any algebraic equation with rational coefficients. This means that π could not exactly satisfy equations of the type $\pi^2 = 10$ or $9\pi^4 - 240\pi^2 + 1,492 = 0$. These are equations involving simple integers with powers of π. The number π can be expressed as an endless continued fraction or as the limit of an infinite series. The remarkable fraction 355/113 expresses π to six decimal places.

2. When I use the word *random* to describe the digits of π, I really mean that they are "normal." (A normal number is defined as a number in which every possible block of digits is equally likely to occur.) The digits of π are clearly not random in the strict sense of the word because if you were to ask for the *n*th digit, you can always find it with absolute certainly by a well-defined procedure. In other words, the "next" digit of π is completely determined, so it is not a random sequence of digits in the strict mathematical sense. Even the term *patternless* is vague, because π obviously has a pattern defined by the formula for it.

3. The probability that any two randomly chosen integers do not have a common divisor is 6 divided by π^2. For example, 4 and 9 do not have a common divisor, but 6 and 9 do (they are both divisible by 3). Why on Earth does π unexpectedly crop up here?

Chapter 9 The Loom of Creation

1. J. Madachy, *Madachy's Mathematical Recreations* (New York: Dover, 1979).

2. C. Pickover, *Chaos in Wonderland: Visual Adventures in a Fractal World* (New York: St. Martin's, 1994).

Chapter 12 The Leviathan Number

1. More recently, mystical individuals of the extreme fundamentalist Right have noted that each word in the name Ronald Wilson Reagan has six letters. Interestingly, the book called The Revelation of St. John the Divine (King James version) has never been accepted in the Syriac-speaking church.

Chapter 14 Fractal Milkshakes and Infinite Archery

1. L. R. Ford was quite modest when he first published on this fascinating geometrical construction. In fact, he begins his article with, "Perhaps the author owes an apology to the reader for asking him to lend his attention to so elementary a subject, for the fractions to be discussed in this paper are, for the most part, the halves, quarters, and thirds of arithmetic."

Chapter 18 Infinitely Exploding Circles

1. E. Kasner and J. Newman, *Mathematics and the Imagination* (Redmond, WA: Tempus Books 1989); (reprint of 1948 book).
2. H. Haber, "Das mathematische Kabinett," *Bild der Wissenshaft*, 2 (April 1964): 73.
3. C. Bouwkamp, "An Infinite Product," *Koninkl. Nederl. Akademi van Wetenschappen—Amsterdam* (1965). Reprint from *Proceedings, Series A,* 68(1), and Indag. Math., 27(1).
4. Ibid.

Chapter 19 The Infinity Worms of Callisto

1. The various worms' log equations come from a tattered pamphlet I found in a garage sale. The pamphlet, published in 1910 by Cambridge University Press, was titled *Orders of Infinity: The Infinitärcalcül of Paul du Bois-Reymond.* The author was G. H. Hardy.

Chapter 22 The Loneliness of the Factorions

1. For a positive integer n, the product of all the positive integers less than or equal to n is called "n factorial," usually denoted as "$n!$". For example, $3! = 3 \times 2 \times 1$.
2. Parenthetically I should point out that Herve Bronnimann from Princeton University has recently found some magnificent factorions in other bases, most notably 519,326,767 which in base 13 is written as 8.3.7.9.0.12.5.11 and is equal to $8! + 3! + 7! + 9! + 0! + 12!$ $+ 5! + 11!$ (You can interpret this base 13 number as $8 \times 13^7 + 3 \times 13^6 + 7 \times 13^5 +$ $9 \times 13^4 + 0 \times 13^3 + 12 \times 13^2 + 5 \times 13^1 + 11 \times 13^0$. Some write this number as $83,790C5B_{13}$). For those of you not familiar with numbers represented in bases other than 10 (which is the standard way of representing numbers), consider how to represent any number in base 2. Numbers in base 2 are called binary numbers. The presence of a 1 in a digit position of a number in base 2 indicates that a corresponding power of 2 is used to determine the value of the binary number. A 0 in the number indicates that a corresponding power of 2 is absent from the binary number. The binary number 1111 repre-

sents $(1 \times 2^3) + (1 \times 2^2) + (1 \times 2^1) + (1 \times 2^0) = 15$. The binary number 1000 represents $1 \times 2^3 = 8$.

Chapter 23 Escape from Fractalia

1. I think of Dr. Sprott as the guru of strange attractors—intricate mathematical shapes that represent the chaotic behavior of physical and mathematical systems. When I asked Dr. Sprott to describe himself, he told me the following. "I'm sure I have the world's largest collection of strange attractors. I estimate about a million. I've looked at maybe 5 percent of them. The others sit in a compact code on my disk. I'm busily doing statistical studies on them. Who am I? B.S. in physics from MIT in 1964, Ph.D. in physics from University of Wisconsin–Madison in 1969. Worked at Oak Ridge National Lab before returning to a faculty position in the Physics Department at the University of Wisconsin–Madison in 1973, where I've been ever since. My training and research have mostly been in experimental plasma physics. I have been heavily involved in physics popularization through a program called 'The Wonders of Physics' that involves public presentations of dramatic demonstrations of physical phenomena, production of videotapes, and development of physics educational software. I'm technical editor of Physics Academic Software. I've written books such as *Introduction to Modern Electronics* (Wiley, 1979) and *Numerical Recipes and Examples in BASIC* (Cambridge University Press, 1991). I became a chaos enthusiast in 1989."

2. J. C. Sprott, *Strange Attractors: Creating Patterns in Chaos* (New York: M&T Books, 1993).

3. C. A. Pickover, *Mazes for the Mind: Computers and the Unexpected* (New York: St. Martin's, 1992).

4. H. O. Peitgen and P. H. Richter, *The Beauty of Fractals: Images of Complex Dynamical Systems* (New York: Springer-Verlag, 1986).

Chapter 25 The Crying of Fractal Batrachion 1,489

1. J. Conway, "Some Crazy Sequences," videotaped talk at AT&T Bell Labs, July 15, 1988.

2. M. Schroeder, *Fractals, Chaos, Power Laws* (New York: Freeman, 1991) .

3. C. Mallows, "Conway's Challenge Sequence," *American Mathematics Monthly* (January 1991): 5–20.

4. D. Hofstadter, *Gödel, Escher, Bach* (New York: Vintage Books, 1980).

5. C. Pickover, "The Drums of Ulupu," in *Mazes for the Mind: Computers and the Unexpected* (New York: St. Martin's, 1993).

Chapter 27 Recursive Worlds

1. B. Mandelbrot, *The Fractal Geometry of Nature* (New York: Freeman, 1982).

2. D. Hilbert, "Über die stetige Abbildung einer Linie auf ein Flachenstück," *Mathematische Annalen*, 38 (1891): 459–460.

3. G. Peano, "Sur une courbe, gui remplit une aire palne," *Mathematische Annalen*, 36 (1890): 157–160.

4. R. Shepard, *Mind Sights* (New York: Freeman, 1991) .

Chapter 30 Vampire Numbers

1. Versions of J for OS/2, Apple, Next, PC, Windows, Sparc, and many other platforms are available via anonymous ftp from watserv1.uwaterloo.ca in languages/apl/j/exec. C source code is also available. See also K. Iverson, *J: Introduction and Dictionary* (Toronto: Iverson Software, 1993). The publisher's address is 33 Major Street, Toronto, Ontario M5S 2K9 Canada.

Chapter 31 Computers, Randomness, Mind, and Infinity

1. M. Schroeder, *Fractals, Chaos, Power Laws* (New York: Freeman, 1991).

2. J. Collins, M. Fanciulli, R. Hohlfeld, D. Finch, G. Sandri, and E. Shtatland, "A Random Number Generator Based on the Logit Transform of the Logistic Variable," *Computers in Physics*, 6(6) (1992): 630–632.

3. S. Ulam and J. von Neumann, "On Combination of Stochastic and Deterministic Processes." *Bulletin of the American Mathematical Society*, 53 (1947): 1120.

4. P. Stein and S. Ulam, "Nonlinear Transformation Studies on Electronic Computers." *Rozprawy matematyczne*, 39 (1964): 1.

5. S. Grossman and S. Thomae, "Invariant Distributions and Stationary Correlation Functions," *Zeitsuhrift für Naturforschung*, 32a (1977): 1353.

6. T. Richards, "Graphical Representation of Pseudorandom Sequences," *Computers and Graphics*, 13(2) (1989): 261–262.

7. D. Knuth, *The Art of Computer Programming, Vol. 2*, 2nd ed. (Reading, MA: Addison-Wesley, 1981).

8. C. Pickover, "Synthesizing Extraterrestrial Terrain," *IEEE Computer Graphics and Applications*, 15(2) (March, 1995): 18–21.

9. J. Colonna, "The Subjectivity of Computers." *Communications of the ACM*, 36(8) (July 1994): 15–18.

10. F. James, "RANLUX: A Fortran Implementation of the High-Quality Pseudorandom Number Generator of Luscher," *Computer Physics Communications*, 79(1) (February 1994): 111–114.

Appendix 2 For Further Exploration

1. Gofer, designed by Mark Jones of Yale, is a very near relation to Haskell (see A. Davie, *An Introduction to Functional Programming Systems Using Haskell*, 1992). In particular, Gofer supports lazy evaluation, higher-order functions, polymorphic typing, pattern matching, support for overloading, etc. (ftp: nebula.cs.yale.edu, directory: pub/haskell/gofer). Gofer runs on a wide range of machines including PCs, Ataris, Amigas, etc., as well as larger UNIX-based systems. A version for the Apple Macintosh has been produced and is available by anonymous ftp from ftp.dcs.glasgow.ac.uk in a subdirectory of pub/haskell/gofer.

Further Reading

Mathematics is the only infinite human activity. It is conceivable that
humanity could eventually learn everything in physics or biology. But
humanity certainly won't ever be able to find out everything in
mathematics, because the subject is infinite. Numbers themselves are
infinite.
— Paul Erdös

General

Berezin, A. (1987). "Super Super Large Numbers." *Journal of Recreational Math,* 19(2): 142–
143.

Ford, L. R. (1938). "Fractions." *American Mathematics Monthly,* 45: 586–601.

Gardner, M. (1983). "Alephs and Supertasks." In *Wheels, Life, and Other Mathematical
Amusements.* New York: Freeman. (In this book Gardner discusses different levels of in-
finity such as $\aleph_0$ and $\aleph_1$ and also "supertasks" involving infinity machines such as the
Thomson lamp, which switches at faster and faster rates.)

Grünbaum, A. (1968). "Are Infinity Machines Paradoxical?" *Science,* 159 (January 26): 396–
406.

Grünbaum, A. (1969). "Can an Infinitude of Operations Be Performed in a Finite Time?" *British Journal for the Philosophy of Science,* 20 (October): 203–218.

Johnson, J. (1994). "Extraordinary Science and the Strange Legacy of Nikola Tesla." *Skeptical
Inquirer* 18(4) (Summer): 366–378.

Kilmister, C. (1980). "Zeno, Aristotle, Weyl and Shuard: Two-and-a-half Millennia of Worries
over Number." *The Mathematical Gazette,* 64 (October): 149–158.

Pennington, J. (1957). "The Red and White Cows." *American Mathematics Monthly,* 64
(March): 197–198.

Pickover, C. (1990). "Is There a Double Smoothly Undulating Integer?" In *Computers, Pattern, Chaos and Beauty.* New York: St. Martin's.

Rademacher, H. (1983). *Higher Mathematics from an Elementary Point of View.* Boston: Birk-
hauser.

Schroeder, M. (1991). *Fractals, Chaos, Power Laws: Minutes from an Infinite Paradise.* New York: Freeman. (Discusses similar sequences, including what Schroeder has called the "rabbit sequence": 1 0 1 1 0 1 0 1 . . . .)

Preface

Gamow, G. (1988). *One, Two, Three... Infinity.* New York: Dover.

Gibilisco, S. (1990). *Reaching for Infinity.* Blue Ridge Summit, Pennsylvania: TAB Books.

Hemmings, R., and Tahta, D. (1984). *Images of Infinity.* Burlington, Vermont: Leapfrogs Insight Series.

Horgan, J. (1994). "Anti-omniscience." *Scientific American*, 271(2) (August): 20–22.

Kasner, E., and Newman, J. (1989). *Mathematics and the Imagination.* Redmond, Washington: Tempus.

Maor, E. (1991). *To Infinity and Beyond.* Princeton, NJ: Princeton University Press.

Rucker, R. (1983). *Infinity and the Mind.* New York: Bantam.

Schroeder, M. (1991). *Fractals, Chaos, Power Laws: Minutes from an Infinite Paradise.* New York: Freeman.

Chapter 21 The Fractal Golden Curlicue Is Cool

Berry, M., and Goldberg, J. (1988). "Renormalization of Curlicues." *Nonlinearity*, 1: 1–26.

Markowsky, G. (1992). "Misconceptions about the Golden Ratio." *College Mathematics Journal*, 23(1) (January): 2–19.

Moore, R., and van der Porrten, A. (1989). "On the Thermodynamics of Curves and Other Curlicues." *McQuarie University Mathematics Reports 89–0031*, April.

Pickover, C. (1993). *Mazes for the Mind: Computers and the Unexpected.* New York: St. Martin's.

Sedgwick, R. (1988). *Algorithms.* New York: Addison-Wesley.

Stewart, I. (1992). *Another Fine Math You've Got Me Into . . .* New York: Freeman.

Chapter 27 Recursive Worlds

Berczi, S. (1986) "Escherian and Non-Escherian Developments of New Frieze Types in Hantis and Old Hungarian Communal Art." In Coxeter, H., ed., *M.C. Escher: Art and Science.* (pp. 349–358). New York: Elsevier.

Berczi, S. (1989). "Symmetry and Technology in Ornamental Art of Old Hungarians and Avar-Onogurians from the Archeological Finds of the Carpathian Basin, Seventh to Tenth A.D." *Computers and Mathematics with Applications*, 17: 715–730.

Berczi, S. (1990). "Local and Global Model of Fibonacci Plant Symmetries." In Gruber, B., and Yopp, J., eds., *Symmetries in Science,* Vol. 4 (pp. 15–28). New York: Plenum.

Hofstadter, D. (1985). *Gödel, Escher, Bach.* Stuttgart: Klett.

Ikeda, K. (1990). "Is a Self-Recursive Biology Possible?" *Rivista di Biologia (Biology Forum)*, 83: 93–106.

Landini, G. (1991). "A Fractal Model for Periodontal Breakdown in Periodontal Disease." *Journal of Periodontal Research,* 26: 176–179.

Pickover, C. (1992). *Mazes for the Mind: Computers and the Unexpected.* New York: St. Martin's.

Pickover, C. (1993). "Recursive Worlds." *Dr. Dobb's Software Journal,* 18(9) (September): 18–26.

Zimmerman, R. E. (1991). "The Anthropic Cosmological Principle: Philosophical Implications of Self-Reference." In Steen, L. A., ed., *Mathematics Today: Twelve Informal Essays* (pp. 15–53). New York: Springer.

Chapter 31 Computers, Randomness, Mind, and Infinity

Clark, R. N. (1985). "A Pseudorandom Number Generator." *Simulation,* 45(5): 252–255.

Ferrenberge, A., Landau, D., and Wong, Y. (1992). "Monte Carlo Simulations: Hidden Errors from "Good" Random Number Generators." *Physical Review Letters* 69: 3382–3384.

Gordon, G. (1978). *System Simulation.* Englewood Cliffs, NJ: Prentice Hall.

Hennecke, M. (1994). "RANEXP: Experimental Random Number Generator Package." *Computer Physics Communications,* 79: 261.

James, F. (1990). "A Review of Pseudorandom Number Generators." *Computer Physics Communications,* 60: 329–344.

Jefferey, H. (1994). "Fractals and Genetics in the Future." In Pickover, C. ed., *Visions of the Future: Art, Technology, and Computing in the 21st Century.* New York: St. Martin's.

Luscher, M. (1994). "A Portable High-Quality Random Number Generator for Lattice Field Theory Simulations." *Computer Physics Communications,* 79(1) (February): 100–110.

Marsaglia, G., and Zaman, A. (1991). "A New Class of Random Number Generators." *Annals of Applied Probability,* 1(3): 462–480.

Marsaglia, G., and Zaman, A. (1994). "Some Portable Very-Long-Period Random Number Generators." *Computers in Physics,* 8(1): 117–121.

Mckean, K. (1987). "The Orderly Pursuit of Disorder." *Discover* (January): 72–81.

Park, S., and Miller, K. (1988). "Random Number Generators: Good Ones Are Hard to Find." *Communications of the ACM,* 31(10): 1192–1201.

Pickover, C. (1990). *Computers, Pattern, Chaos and Beauty.* New York: St. Martin's.

Pickover, C. (1991). *Computers and the Imagination.* New York: St. Martin's.

Pickover, C. (1991). "Picturing Randomness on a Graphics Supercomputer." *IBM Journal of Research and Development,* 35(1/2) (January/March): 227–230.

Pickover, C. (1992). *Mazes for the Mind: Computers and the Unexpected.* New York: St. Martin's.

Pickover, C. (1994). *Chaos in Wonderland: Visual Adventures in a Fractal World.* New York: St. Martin's.

Pickover, C. (1995). "Random Number Generators: Pretty Good Ones Are Easy to Find." *Visual Computer,* in press.

Pickover, C. and Tewksbury, S. (1994). *Frontiers of Scientific Visualization.* New York: Wiley.

Schroeder, M. (1990). "Random Number Generators." In *Number Theory in Science and Communication,* 2nd enlarged ed. (pp. 289–295). Berlin and New York: Springer.

Voelcker, J. (1988). "Picturing Randomness." *IEEE Spectrum,* 25(8) (August): 13.

Index

About the Author

Clifford A. Pickover received his Ph.D. from Yale University's Department of Molecular Biophysics and Biochemistry. He graduated first in his class from Franklin and Marshall College, after completing the four-year undergraduate program in three years. He is author of the popular books *Chaos in Wonderland: Visual Adventures in a Fractal World* (1994), *Mazes for the Mind: Computers and the Unexpected* (1992), *Computers and the Imagination* (1991), and *Computers, Pattern, Chaos, and Beauty* (1990)—all published by St. Martin's Press—as well as the author of more than 200 articles concerning topics in science, art, and mathematics. He is also coauthor, with Piers Anthony, of the science-fiction novel *Spider Legs*.

Pickover is currently an associate editor for the scientific journals *Computers and Graphics* and *Computers in Physics*, and is an editorial board member for *Speculations in Science and Technology, Idealistic Studies, Leonardo*, and *YLEM*. He has been a guest editor for several scientific journals. He is editor of *The Pattern Book: Fractals, Art, and Nature* (World Scientific, 1995), *Visions of the Future: Art, Technology, and Computing in the Next Century* (St. Martin's, 1993), and *Visualizing Biological Information* (World Scientific, 1995) and coeditor of the books *Spiral Symmetry* (World Scientific, 1992) and *Frontiers in Scientific Visualization* (Wiley, 1994). Dr. Pickover's primary interest is in scientific visualization.

In 1990, he received first prize in the Institute of Physics' Beauty of Physics photographic competition. His computer graphics have been featured on the cover of many popular magazines, and his research has recently received considerable attention by the press—including CNN's *Science and Technology Week, Science News, The Washington Post, Wired*, and *The Christian Science Monitor*—and also in international exhibitions and museums. *OMNI* magazine recently described him as "Van Leeuwenhoek's twentieth-century equivalent." The July 1989 issue of *Scientific American* featured his graphic work, calling it "strange and beautiful, stunningly realistic." Pickover has received U.S. Patent 5,095,302 for a 3-D computer mouse. Dr. Pickover is also a consultant for WNET on science education projects, a novelist, and a frequent columnist for *Discover* magazine.